Me++

Me++

THE CYBORG SELF AND THE NETWORKED CITY

WILLIAM J. MITCHELL

The MIT Press   Cambridge, Massachusetts   London, England

This book was set in Garamond 3, Magda Clean Mono, and Magda Clean by SNP Best-set Typesetter Ltd., Hong Kong.

Printed and bound in the United States of America.

Library of Congress Cataloging-in-Publication Data

Mitchell, William J. (William John), 1944–
  Me++ : the cyborg self and the networked city / William J. Mitchell.
    p.   cm.
  ISBN 0-262-13434-9 (hc : alk. paper)
  1. Telecommunication—Popular works. 2. Computer networks—Popular works. 3. Cyberspace—Popular works. 4. Information superhighway—Social aspects. I. Title.

TK5101.M555 2003
303.48′3—dc21

                                        2003053968

Radio is one-sided when it should be two-. It is purely an apparatus for distribution, for mere sharing out. So here is a positive suggestion: change this apparatus over from distribution to communication. The radio would be the finest possible communication apparatus in public life, a vast network of pipes. That is to say, it would be if it knew how to receive as well as to transmit, how to let the listener speak as well as hear, how to bring him into a relationship instead of isolating him. On this principle the radio should step out of the supply business and organize its listeners as suppliers.

Bertolt Brecht, "The Radio as an Apparatus of Communication," 1926

CONTENTS

Me++

PROLOGUE

To honor an approaching centenary, I embarked on an electronic pilgrimage. Like Ulysses setting sail for Ithaca, I looked to the heavens for guidance.[1] I entered "Marconi Station, Wellfleet, Massachusetts" into my automobile navigation system. GPS satellites located me, software computed the quickest route, and synthetic voice commands conducted me to a sandy cliff on the shores of Cape Cod.

The motorists whizzing past on Route 6, listening to their radios and casually chatting on their cellphones, didn't give it a thought, but it was on this windy site (most of which has now fallen into the ocean) that the wireless world began.[2] Here, Guglielmo Marconi built four 210-foot towers, spun a spiderweb of wires in the sky, cranked up a kerosene engine to drive a 20,000-volt power supply, and ran a spark-gap rotor that could be heard for miles. On 18 January 1903 (the year in which the Wright brothers were to introduce the world to powered flight), he transmitted a wireless telegraph message across the Atlantic. Suddenly the continents were connected in a new way.

A century later, global wireless systems brought me to the spot and kept me effortlessly in touch as I stood there. In my hand I held an inexpensive transmitter and receiver that was immeasurably more sophisticated than Marconi's immense construction, and could instantly connect me to any one of hundreds of millions of similar devices scattered around the world. Furthermore, it could link me to the countless servers of the Internet and the Web. I pulled off the cover (no doubt voiding the warranty) to reveal a palm-sized, precisely made architectural model; the powerhouse had shrunk to a matchbook-scaled battery, the transmission house now resided on a chip, and the antenna tower was just a couple of inches long. It seemed that Brobdingnag had been rebooted as Lilliput—producing a dramatic inversion. Where Marconi's human operators had been appendages to a motionless machine, the device that I now grasped was a liberating extension of my mobile body.

SCALE

The two parts of Marconi's system had evolved in opposite directions. The network had scaled up. The single wireless link had exploded into a dense, global web of wireless infrastructure; if you counted all of its terrestrial, satellite, and spacecraft linkages, it was now humankind's most extensive single construction. Simultaneously, though, the transmission and reception apparatus had dramatically scaled down; it had reduced from landscape element to fashion accessory.

And this two-sided transformation, I reflected, was not just a technological wonder; it had changed the way we lived. Over the course of a hundred years—and particularly in the last couple of decades—the convergence of increasingly capable wireless technology with expanding network infrastructure, miniaturized electronics, and proliferating digital information had radically refashioned the relationships of individuals to their constructed environments and to one another. I thought of the cellphone conversations that had continued until the very moment the World Trade Center towers collapsed, of the desperate calls from the cabins of hijacked airliners hurtling toward their targets, and of the pagers that had carried on transmitting from beneath the rubble.[3] We had become inseparable from our increasingly capable electronic organs; our very limbs had become fleshy antenna supports; our interconnections had ramified and intensified to an almost incomprehensible degree. From prenatal imaging and heartbeat monitoring to posthumous persistence of our digital addresses and traces, our bodies now existed in a state of continuous electronic engagement with their surroundings.

Thing-to-thing relationships had similarly mutated. Digital networks had begun as collections of large, expensive boxes connected by cheap wires. But over time the boxes got smaller, cheaper, and more numerous, while the wires remained much the same. Increasingly, then, it was the wiring of networks, not the boxes, that consumed space and cost money. With the extension of networks, telecommunications providers worried about the "last mile" of cable to homes and businesses, architects struggled to jam cable trays and wiring closets into buildings, cables snaggled across floors in unseemly tangles, and many things that might benefit from network connections never got them because it was just too difficult to run the wires. By the dawn of the twenty-first century, though, inexpensive, ubiquitous *wireless* connections were linking whole new classes of things into networks— very tiny things, very numerous things, very isolated things, highly mobile things, things deeply embedded in other things, and things that were jammed into tight and inaccessible places. Most dramatically, wireless transponders that could identify physical objects had shrunk to the size of a pinhead and the cost of a few cents, and billions of them were being produced and deployed. Now, nothing need be without processing power, and nothing need be left unlinked. The distinction between computer hardware and more traditional sorts of hardware was rapidly fading.

MATERIALITY

The trial separation of bits and atoms is now over. In the early days of the digital revolution it seemed useful to pry these elementary units of materiality and information apart. The virtual and the physical were imagined as separate realms—cyberspace and meatspace, as William Gibson's insouciantly in-your-face formulation put it. This seemed a welcome release from the stubborn constraints of tangibility— until the dot-com bubble burst, at least. Now, though, the boundary between them is dissolving. Networked intelligence is being embedded everywhere, in every kind of physical system—both natural and artificial. Routinely, events in cyberspace are being reflected in physical space, and vice versa.[4] Electronic commerce is not, as it turns out, the replacement of bricks and mortar by servers and telecommunications, but the sophisticated integration of digital networks with physical supply chains. Increasingly, we are living our lives at the points where electronic information flows, mobile bodies, and physical places intersect in particularly useful and engaging ways. These points are becoming the occasions for a characteristic new architecture of the twenty-first century.

The metaphor of "virtuality" seemed a powerful one as we first struggled to understand the implications of digital information, but it has long outlived its usefulness. Bits don't just sit out there in cyberspace, to be visited occasionally like pictures in a gallery or poked at through a "window" (telling metaphor). It makes more sense to recognize that invisible, intangible, electromagnetically encoded information establishes new types of relationships among *physical* events occurring in *physical* places. A bit is indeed a "difference that makes a difference," but we should think of that difference as something concrete, with definite spatial and temporal coordinates—such as the opening or closing of some particular switch, door, or floodgate, the dispatch of a product from a warehouse, the starting or stopping of a motor, the change in color of a pixel on a screen somewhere, or the shift in position of a robot arm—rather than a change in value of an abstract variable or an alteration in the imagined mental state of some idealized message recipient. Bits organized into code now constitute the most powerful means we have for expressing intentions and translating them into actions.

INTENTIONS

Conceived of in this tangible and architectural fashion, code makes it possible to specify the relationships between physical causes and effects in terms of symbolically expressed mathematical functions, rather than by constructing specialized mechanical linkages or hydraulic systems, hardwired electrical circuits, and the like. Through wired and wireless telecommunication, code makes these relationships effective at a distance. Through its capacity to store instructions and data, code allows predefined relationships to manifest themselves asynchronously. Through the increasing integration of telecommunication networks and digital controls with vehicles and transportation networks, electrical supply systems, water, gas, and petroleum pipeline systems, dams and flood control systems, air-conditioning systems, and global trading systems, digital code now controls the supply of just about everything that's essential to us. Code is weaving an ever-denser web of complex, inescapable interconnections across space and time. And this is just the beginning; the curve of technological development is snapping into the steep part.

But there are bad bits as well as good ones. Ill-intended zeros and ones can serve as weapons as surely as bullets. In a networked, electronically interconnected world, there is no fundamental distinction between addresses and targets. My conventional mail address makes me a potential target for junk mail, anthrax spores, and letter bombs, dispatched from just about anywhere. Similarly, my electronic mail address attracts viruses, worms, and spam—of even less determinate origin. The coordinates propagated by my wireless devices invite electronic tracking and surveillance. The technology that so precisely guides my automobile to a specified urban location can also guide a missile or a smart bomb. And increasingly, the containers that now speed through the links we have so carefully constructed—from packets on the Internet to the cabins of jetliners—are not only subject to physical hijacking but also to electronic hacking, militarization, reengineering, reprogramming, and being recombined, redirected, and turned against us.

ETHICS

In the following pages I shall examine the interwoven implications of wireless linkage, ubiquitous and global-scale interconnection, miniaturization and portability, mobilized bits, and associated systems and practices for our bodies, our clothing, our architecture, our cities, our patterns and systems of movement, and our uses of space and time. In particular, I shall consider a crucial consequence of these technological transformations—the shift from a world structured by boundaries and enclosures to a world increasingly dominated, at every scale, by connections, networks, and flows. This is a world of less rigid, more fluid and flexible relationships—of knowledge to action, of shape to materials, and of people to places.

It is a world, I shall argue, in which networks propagate the effects of our actions far beyond traditional boundaries. Increasingly, we can do unto others at a distance, and they can do unto us. The widening circle of electronic interconnectivity and interdependence creates a widening moral circle. The principle of reciprocity—the ancient Golden Rule—is no longer circumscribed in space and time, and its expansion has profound consequences for design, engineering, and planning practice.

My analysis of this condition gives priority neither to the autonomous logic of technological development nor to desire and power. Instead, I assume that we shape our technologies, then our technologies shape us, in ongoing cycles that produce our everyday physical and social environments. I write not from the perspective of a technological futurist who attempts to predict these cycles, nor that of an empirical social scientist who carefully observes them unfolding, but that of a critically engaged designer whose business it is to reflect, imagine, and invent.

1

BOUNDARIES/NETWORKS

Consider, if you will, Me++.

    I consist of a biological core surrounded by extended, constructed systems of boundaries and networks. These boundary and network structures are topological and functional duals of each other.[1] The boundaries define a space of containers and places (the traditional domain of architecture), while the networks establish a space of links and flows. Walls, fences, and skins divide; paths, pipes, and wires connect.

BOUNDARIES

My natural skin is just layer zero of a nested boundary structure. When I shave, I coat my face with lather. When I'm nearly naked in the open air, I wear—at the very least—a second skin of spf 15 sunblock.

    My clothing is a layer of soft architecture, shrinkwrapped around the contours of my body. Beds, rugs, and curtains are looser assemblages of surrounding fabric—somewhere between underwear and walls. My room is a sloughed-off carapace, cast into a more rigorous geometry, fixed in place, and enlarged in scale so that it encloses me at a comfortable distance. The building that contains it has a weatherproof exterior shell. Before modern mobile artillery, fortified city walls would have provided a final, hardened, outermost crust; these sorts of urban-scale skins remained reasonably effective at least until the 1871 siege of Paris, during the Franco-Prussian War.[2]

    In the early years of the Cold War, outer defensive encasements reemerged, in extreme form, as domestic nuclear bunkers. The destruction of the Berlin Wall in 1989 marked the end of that edgy era. But still, if I end up in jail, an internment camp, or a walled retirement community, the distinction between intramural and extramural remains brutally literal. If I retire to a farm, a boundary fence stops my stock from straying. And if I locate myself within the homeland of a major military power, I take refuge behind a dubious high-tech bulwark that extends across thousands of kilometers; our extradermal armored layers have coevolved, with increasingly fearsome weapons systems, into invisible radar curtains and missile shields that create vast electronic enceintes. I surround myself with successive artificial skins that continually vary in number and character according to my changing needs and circumstances.[3]

All of my boundaries depend, for their effectiveness, upon combining sufficient capacity to attenuate flow with sufficient thickness. If I want to keep warm, for example, I can use a thin layer of highly insulating material or a thicker layer of a less effective insulator. If I want acoustic privacy, I can retreat behind a closed door, or I can simply rely on the attenuation of sound waves in air and move out of earshot. If I want to create a jail, I can construct escape-proof walls, or I can remove the prisoners to a sufficiently distant place—like the eighteenth-century British convicts transported to Australia. In sparsely populated territories, distance creates many natural barriers, while in buildings and cities, efficient artificial barriers subdivide closely packed spaces.

CONNECTIONS

But I am, as Georg Simmel observed, a "connecting creature who must always separate and who cannot connect without separating."[4] My enclosures are leaky. Crossing the various boundaries that surround me there are paths, pipes, wires, and other channels that spatially concentrate inflows and outflows of people, other living creatures, discrete goods, gases and fluids, energy, information, and money. I am inextricably entangled in the networks of my air, water, waste disposal, energy, transportation, and Internet service providers.

To create and maintain differences between the interiors and exteriors of enclosures—and there is no point to boundaries and enclosures if there are no differences—I seek to control these networked flows. So the crossing points are sites where I can survey what's coming and going, make access decisions, filter out what I don't want to admit or release, express desire, exercise power, and define otherness. Directly and indirectly, I employ doors, windows, bug screens, gates, cattle grids, adjustable apertures, valves, filters, prophylactics, diapers, face masks, receptionists, security checkpoints, customs and immigration checkpoints, traffic signals, routers and switches to determine who or what can go where, and when they can go there. So do you, of course, and so do others with the capacity to do so in particular contexts.

Through the interaction of our efforts to effect and control transfers among enclosures and our competition for network resources, we mutually construct and constrain one another's realms of daily action. Within the relatively stable framework of our interconnecting, overlapping, sometimes shared transfer networks, our intricately interwoven demands and responses create fluctuating conditions of freedom and constraint. And as networks become faster, more pervasive, and more essential, these dynamics become increasingly crucial to the conduct of our lives; we have all discovered that a traffic jam, a check-in line, a power outage, a server overwhelmed by a denial-of-service attack, or a market crash can create as effective a barrier as a locked door. The more we depend upon networks, the more tightly and dynamically interwoven our destinies become.

NETWORKS

The archetypal structure of the network, with its accumulation and habitation sites, links, dynamic flow patterns, interdependencies, and control points, is now repeated at every scale from that of neural networks (neurons, axons, synapses) and digital circuitry (registers, electron pathways, switches) to that of global transportation networks (warehouses, shipping and air routes, ports of entry).[5] And networks of different types and scales are integrated into larger network complexes serving multiple functions. Depending upon our relationships to the associated social and political structures, each of us can potentially play many different roles (some strong, some weak) at nodes within these complexes—owner, authorized user, operator, occupant, occupier, tenant, customer, guest, sojourner, tourist, immigrant, alien, interloper, infiltrator, trespasser, snooper, besieger, cracker, hijacker, invader, gatekeeper, jailer, or prisoner. Power and political identity have become inseparable from these roles.

With the proliferation of networks and our increasing dependence upon them, there has been a gradual inversion of the relationship between barriers and links. As the ancient use of a circle of walls to serve as the ideogram for a city illustrates, the enclosing, dividing, and sometimes-defended *boundary* was once the decisive mechanism of political geography. Joshua got access the old-fashioned way; when he blew his righteous trumpet, the walls of Jericho came tumbling down. By the mid-twentieth century, though, the most memorable ideogram of London was its underground network, and that of Los Angeles was its freeway map; riding the networks, not dwelling within walls, was what made you a Londoner or an Angeleno. And the story of recent urban growth has not been one of successive encircling walls, as it mostly would have been for ancient, medieval, and Renaissance cities, but of network-induced sprawl at the fringes.

More recently, the unbelievably intricate diagram of Internet interconnectivity has become the most vivid icon of globalization. Now you get access by typing in your password, and IT managers dissolve the perimeters between organizations by merging their network access authorization lists. Today the *network*, rather than the enclosure, is emerging as the desired and contested object: the dual now dominates.[6] Extension and entanglement trump enclosure and autonomy. Control of territory means little unless you also control the channel capacity and access points that service it.

A year after the September 11 attacks on New York and Washington, the implications of this were sinking in. The President's Critical Infrastructure Protection Board bluntly reported (to nobody's very great surprise),

> Our economy and national security are fully dependent upon information technology and the information infrastructure. A network of networks directly supports the operation of all sectors of our economy—energy (electric power, oil and gas), transportation (rail, air, merchant marine), finance and banking, information and telecom-

munications, public health, emergency services, water, chemical, defense industrial base, food, agriculture, and postal and shipping. The reach of these computer networks exceeds the bounds of cyberspace. They also control physical objects such as electrical transformers, trains, pipeline pumps, chemical vats, radars, and stock markets.[7]

Connectivity had become the defining characteristic of our twenty-first-century urban condition.

CLOCKS

All networks have their particular paces and rhythms. Within the nested layers and recursively embedded networks of my world, my pulse—the sound of an intermediate-scale, low-speed vascular network—has been mechanized, regularized, externalized, and endlessly echoed back to me. Just as boundary, flow, and control systems subdivide my space into specialized, manageable zones, these constructed rhythms partition my time into discrete, identifiable, assignable, sometimes chargeable chunks. Bean counters are also minute counters; measurable, accountable time is money.

The miraculously monotone beat of the pendulum first established this possibility.[8] Ancient sundials and water clocks had marked the flow of time, and Benedictine monastery bells had formalized its approximate mechanical subdivision. Clock towers had provided European towns with faster communal heartbeats—essential, as Lewis Mumford pointed out, to the regulation and coordination of social and economic life, and eventually to the industrial organization of production.[9] Then, in the seventeenth century, Christiaan Huygens devised a pendulum clock that ticked precisely.

This innovation also initiated a shift in scale. Furniture-sized towers (grandfather clocks, standing in domestic hallways) soon began to associate timekeeping with the dwelling and the family rather than with the town and the larger community. Substituting spring-driven mechanisms for pendulums allowed clocks to become even smaller, more portable, and eventually wearable—now associating timekeeping with the individual.[10] Timepieces moved to pockets, then to wrists—provocatively, the organic pulse's most obvious point of presence. Clinging tightly to flesh, they have enabled the large-scale scheduling and coordination of individual activities; during the American Civil War, for example, the Union forces depended upon them to synchronize operations.

As artificial pulse rates have accelerated, timekeeping mechanisms have continued to shrink. Today, the gigahertz, crystal oscillator hearts of tiny computer chips are embedded everywhere. (Chips without clocks are possible, and may turn out to have some important advantages, but they are not yet in widespread use.)[11] Electronic vibrations subdivide seconds into billions of parts, pace the execution of computational tasks, discipline our interactions with computational devices, calibrate GPS navigation systems, regulate power distribution and telephone systems, measure and commodify both human and machine work, and precisely construct the accelerating tempos and rhythms of the digital era—coordinated, where necessary, by a central atomic clock.[12] They not only *mark* time, they *trigger* the execution of instructions and programs. Seconds, milliseconds, microseconds, nanoseconds, picoseconds: the electronic global heartbeat keeps quickening and gathering power—so much so that, when its coordinated microrhythms threatened to falter at Y2K, there was bug-eyed panic in the technochattering classes.[13] There was talk of "spectacular explosions, nuclear meltdowns, power blackouts, toxic leaks, plane crashes, and bank failures."[14]

PROCESSES

But there is, of course, more to the construction of time than the increasingly precise subdivision of the day. As clocks multiply and distribute themselves spatially, the relationships among them begin to matter.

Different places may simply run on their own clocks, or their timekeeping systems may be standardized and synchronized. When there was little communication between spatially separated settlements, local time sufficed, and there was no need for such coordination, but linkage by long-distance railroad and telegraph networks eventually made it imperative. In 1851 the Harvard College

Observatory began to distribute clock ticks, by telegraph, to the railroad companies. As transportation and telecommunication capacities have increased, we have entered the era of globalized network time—of GMT, time zones, and sleep cycles decoupled from the solar day.[15] Once, villagers rose with the roosters to work until sunset in nearby fields; now, jet-lagged business travelers do their email at three a.m. in hotel rooms far from home.

Computers have added additional layers of complexity to the construction of time. The first computers—constructed according to the elegant principles of Turing and von Neumann—were strictly sequential machines, executing one operation at a time; programming was a matter of specifying these operations in precise order. Everything was rigorously governed by clock increments and finite (though small) durations. But as interactive computing developed, a distinction emerged between tasks that could be performed in "real" time and those that could not. For example, computer animations of three-dimensional environments could be computed and stored for later playback, or (as in today's video games) they could be computed and presented on the fly, with no perceptible time lag. In other words, if you take advantage of fast machines to compress processes, you can elide the distinction between simultaneity and sequence. "Virtual reality" would be impossible without this.

The practice of timesharing has produced a further elision. If a processor is fast enough, it can be programmed to divide its time among multiple simultaneous processes—providing the illusion that it is devoting itself exclusively to each one. In effect, a single, sequential processor divides itself into multiple "virtual machines" that seem to occupy the same space and time. The ancient, seemingly unproblematic concept of *hic et nunc*—what's here and now—begins to frazzle.

As processors have become smaller and cheaper, and as they have been integrated into networks, it has become increasingly feasible to program parallel rather than strictly sequential processes; tasks are divided up among multiple processors, which simultaneously contribute to producing the desired result. It is even possible to imagine organizing the entire Internet as a parallel computation device.[16] At this point—particularly as network speeds approach the internal bus speeds of computers—it no longer makes sense to think of a computer as a compact, discrete object, or to distinguish between computers and networks. Eventually, we will approach the physical speed limit, and its associated paradox; information cannot travel faster than light, so spatially distributed events that seem simultaneous from one node in a lightspeed network may seem sequential from another, and vice versa.

The logical endpoint of this shift to networked parallelism is the emerging possibility of quantum computing—in which every atom stores a bit, vast numbers of atomic-scale processing elements are harnessed to execute computations at unprecedented speed, and the notoriously strange spatial and temporal logic of quantum mechanics (rather than the familiar logic of our everyday world) takes over.[17] (It isn't easy to wrap your mind around the fact of quantum systems occupying several places at once, quantum bits registering 0 and 1 at the same time, and quantum computers performing large numbers of computations simultaneously.) And, maybe, the ultimate network will operate by the quantum-magical means of quantum entanglement and teleportation of quantum states from one site to another.[18]

So we have gone from local habitation and mechanical subdivision of time to a far more dynamic, electronically based, network-mediated, global system of sequencing and coordination. The early moderns measured out their lives in clock ticks (and sometimes, as Prufrock lamented, coffee spoons); now, our webs of extension and interconnection run on nanosecond-paced machine cycles that are edging into the domain of quantum logic. The more we interrelate events and processes across space, the more simultaneity dominates succession; time no longer presents itself as one damn thing after another, but as a structure of multiple, parallel, sometimes cross-connected and interwoven, spatially distributed processes that cascade around the world through networks. Once there was a time and a place for everything; today, things are increasingly smeared across multiple sites and moments in complex and often indeterminate ways.

DISCONTINUITIES

In the fast-paced, digitally mediated world that we have constructed for ourselves, what exists between 0 and 1, a pixel and its neighbor, or a discrete time interval and the next? The answer, of course, is nothing—profoundly nothing; there's no there there. The digital world is logically, spatially, and temporally discontinuous.

Our networks are similarly discontinuous structures; they have well-defined access points, and between these points things are in a kind of limbo. If you drop a letter into a mailbox, it disappears into the mail network until it shows up at the recipient's box, and if you send an email, it's just packets in the Internet cloud until it is reassembled upon receipt. Obviously it is possible, in principle, to precisely track things through networks, but in practice we rarely care about this. We experience networks at their interfaces, and only worry about the plumbing behind the interfaces when something goes wrong.

If you transfer *yourself* through a network, you directly experience this limbo. It is, perhaps, most dramatic on intercontinental night flights. You have your headphones on, there is darkness all around, and there is no sensation of motion. The video monitor constructs a local reality, and occasionally interrupts it to display current times at origin and destination. It is best not to worry too much about how to set your watch right now, precisely where you are, or whose laws might apply to you.

The discontinuities produced by networks result from the drive for efficiency, safety, and security. Engineers want to limit the number of access points and provide fast, uninterrupted transfers among these points. So you can drink from a stream anywhere along its length, but you can only access piped water at a faucet. You can pause wherever you want when you're strolling along a dirt track, but you must use stations for trains, entry and exit ramps for freeways, and airports for airline networks—and your experience of the terrain between these points is very limited. You experience the architectural transitions between floors of a building when you climb the stairs, but you go into architectural limbo between the opening and closing of the doors when you use the elevator.

HABITATS

Decades ago, at the very dawn of the digital era, Charles Moore (the most thoughtful architect of emerging postmodernity) shrewdly understood what the simultaneous conditions of extension and discontinuity meant for our daily use of space; our habitats no longer consist of single or contiguous enclosures, but have become increasingly fragmented and dispersed. They are no longer bounded by walls, but by the reach of our networks. They are occupied by spatially dispersed organizations, ranging from multinational corporations and retail chains to terrorist networks. They are controlled and defended not at a continuous perimeter, but at separated and scattered access nodes. They are given order and meaning not by participation in strict spatial sequences and hierarchies, but by their global linkages. Our domains of knowledge and action cannot be defined as fixed neighborhoods, but must now be understood as dynamic, emergent, geographically and temporally fluctuating patterns of presence. In his influential essay "Plug It In, Rameses," he observed:

> The most powerful and effective places that our forebears made for themselves, and left for us, exist in contiguous space. They work on an organized hierarchy of importances, first dividing what is inside from what is outside, then in some way arranging things in order of their importance, so that objects give order to a location, and location gives importance to objects, as at Peking, where an axis penetrates from outside through layer after layer of increasing importance (like the skins of an onion) to the seat of the emperor himself, or as in Hindu towns where caste determined location from clean to dirty along the flow of water which served everyone. . . . Our own places, however, like our lives, are not bound up in one contiguous space. Our order is not made in one discrete inside neatly separated from a hostile outside. . . . We have, as we all know, instant anywhere, as we enjoy our capacity to make immediate electronic contact with people anywhere on the face of the globe. . . . Our new places, that is, are given form with electronic, not visual glue.[19]

COMMUNITIES

Sociologists would use more technical language to make much the same point as Moore's. They would say that I—like most urbanites today—get companionship, aid, support, and social control from a few strong social ties and many weak ones.[20] These ties, which might manifest themselves, for example, as the entries in my cellphone and email directories, establish social networks. In the past, such networks would mostly have been maintained by face-to-face contact within a contiguous locality—a compact, place-based community.[21] Today, they are maintained through a complex mix of local face-to-face interactions, travel, mail systems, synchronous electronic contact through telephones and video links, and asynchronous electronic contact through email and similar media.[22] They are far less dense, and they extend around the world, coming to earth at multiple, scattered, and unstable locations.[23] As Barry Wellman has crisply summarized, "People in networked societies live and work in multiple sets of overlapping relationships, cycling among different networks. Many of the people and the related social networks they deal with are sparsely knit, or physically dispersed and do not know one another."[24]

In the years since Moore wrote, our physical habitats have grown more fragmented and dispersed as transportation networks have extended further and operated faster. Simultaneously, the electronic glue has grown much stronger; it now includes voice, video, and data channels, broadcast and point-to-point links, place-to-place and person-to-person communication, the fixed infrastructure of the bank ATM system, the sleek portable equipment of the corporate road warrior jetting between global cities, and the cheap phone card of the migrant worker.

Wherever I currently happen to find myself, I can now discover many of the same channels on a nearby television, I can access the same bank account, and I can chat with the same people on my cellphone. I can download my email and send replies almost completely independently of location. And my online world, which once consisted of ephemeral and disconnected fragments, has become increasingly persistent, interconnected, and unified; it's there again, pretty much as I left it, whenever I log in again from a new location. The constants in my world are no longer provided by a contiguous home turf: increasingly, my sense of continuity and belonging derives from being electronically networked to the widely scattered people and places I care about.

2

CONNECTING CREATURES

You can read *Ulysses* as *Honey, I Shrunk the City*. Dublin's buildings map to vital organs—the newspaper to the lungs, the concert room to the ear, and so on—while Leopold Bloom and Stephen Dedalus circulate through the urban anatomy like sentient corpuscles, disclosing associated aspects of their biologically embodied consciousnesses at each successive location. Conversely, you can read *Finnegans Wake* as *Attack of the Fifty-Mile Man*. The sleeping body of innkeeper/city-builder H. C. Earwicker (or HCE, Howth Castle and Environs, and Here Comes Everybody) becomes one with Dublin's civic geography; his head is "the macroborg of Holdhard" (the hill of Howth Head), he stretches to the scandalously tumescent "microbirg of Pied de Poudre" (the Powder Magazine in Phoenix Park), and Dubliners are to be found "hopping round his middle like kippers on a griddle." Earwicker/Environs/Everybody's voice is the murmur of Dublin reflecting upon itself.

Now the body/city metaphors have turned concrete and literal. Embedded within a vast structure of nested boundaries and ramifying networks, my muscular and skeletal, physiological, and nervous systems have been artificially augmented and expanded. My reach extends indefinitely and interacts with the similarly extended reaches of others to produce a global system of transfer, actuation, sensing, and control. My biological body meshes with the city; the city itself has become not only the domain of my networked cognitive system, but also—and crucially—the spatial and material embodiment of that system.

Extra muscle first came from animals; horses and riders had all-terrain capability and great power-to-weight ratio. Walking sticks provided an early, rudimentary form of exoskeletal support. Introduction of some elementary mechanisms—wheels, beams, and containers assembled to form carts, together with smoothly paved surfaces—yielded powered vehicles and initiated the long symbiosis of vehicles, roads, and cities.

The first roads were a primitive form of network infrastructure. Wheels were devices operating upon that infrastructure, and each was precisely adapted to the requirements of the other. This marked the beginning of a profoundly consequential coevolutionary process—one in which devices, systems, vehicles, and buildings have adapted themselves to the affordances of available infrastructures, while infrastructures have evolved and multiplied in response to growing demands. The hoof on the dirt track had extended, transformed, and amplified the actions of feet in the stirrups and hands on the reins. Eventually the tire on the road did the same—via mechanical and electronic linkages—for feet on the pedals and hands on the wheel. Through this sort of process, over the centuries, our limbs and muscles have continually extended and elaborated themselves.[1]

In the early industrial era, the steam engine and the horseless carriage substituted machine power for animal muscle. At first, machine power could be transmitted only over short distances, by means of mechanical linkages such as the drive trains of automobiles and the belt systems of early mills, but the combination of electrical generators, transmission systems, and motors allowed cities to equip themselves with extensive, efficient power distribution networks.[2] And as power grids have extended and linked with one another, power sources have increasingly distanced themselves from sites of consumption; Hoover Dam (near Las Vegas) exports most of its hydroelectric power output to southern California, and Hydro-Quebec supplies much of eastern North America. The electric power that is essential to modern cities has become a dynamically priced, computer-controlled commodity that is switched around through vast networks as fluctuations in supply and demand require—and as energy traders determine.

Through the nineteenth and twentieth centuries, by a process of exuberant invention, mechanization increasingly took command.[3] Today, as a privileged postmodern urbanite, I can take advantage of the resulting vast accumulation of mechanical devices to precisely apply machine power wherever and whenever I may need it, with instruments ranging in size from microscopic actuators to hand tools, appliances, vehicles, elevators and escalators, cranes, and conveyor belts, to huge industrial plants. If I operate a telerobot over the Internet, I can extend my grasp and manipulative capacity by thousands of miles. If I have the skills, I can perform telerobotic surgery on a patient on the far side of an ocean.[4] I can even tend a distant garden electronically.[5]

Where a sword might once have lengthened and hardened my hand as a weapon, I could now (as every competent terrorist knows) remotely detonate a bomb simply by attaching a cellphone to it.[6] But that is just the informal violence sector's ad hoc alternative to putting flesh directly on the line. The vast weapon systems of twenty-first-century military organizations are fiendishly extended, multiplied, and strengthened versions of the ancient soldier's legs (which have become military vehicles and delivery systems), sword hand for offense, shield hand for defense, and eyes and ears for intelligence gathering. Since wireless remote control replaced the direct grip of the hand on the weapon, and since cybernetic mechanisms were introduced to control weapon systems more precisely, electronics, software, and robotic mechanisms have increasingly taken over the action.[7] If I serve as an up-to-date military functionary, I am simply (in Norbert Wiener's prescient words) "coupled into the fire-control system and acting as an essential part of it." I become a squishy control node in an extensive and highly integrated machine network.[8] And this condition is generalizing from fire control to choreography of the machines that pervade our daily life.

By programming robotic devices I can precisely specify their future actions. I can instruct them to repeat the same actions indefinitely, to take action at a specified moment (as when a virus wakes up at midnight on a given date and wipes out your hard disk), and to respond to different conditions in different ways. And by copying programs and distributing them to multiple devices, I can repeat the same choreography at different locations. Through electronic storage and distribution of my encoded commands—particularly by means of digital networks—I can indefinitely multiply and distribute my points of physical agency through space and time.

FLOWS (CHANNELED)

Water supply and sewer networks have become geographic extensions of my alimentary canal, my respiratory system, and associated organic plumbing. The carbon-based systems that circulate solids, fluids, and gases within my bag of skin are connected to a vast, external, mostly metallic and plastic network of pipes, ducts, pumps, processing plants, and mechanical transportation devices for food, water, conditioned air, and waste disposal.[9] These extended networks collect resources in distant and dispersed catchment zones, concentrate them at storage nodes, transfer them to consumption nodes, and eventually disperse waste to disposal zones. They enable me to extend my ecological footprint (that is, the land area required to support me and assimilate my waste products) far beyond the scale that was possible before the development of extended plumbing networks—and, indeed, far beyond the point of prudence.

Under the standard arrangement, extended plumbing systems link me into the planet's natural air flow systems, water systems, and food webs, but their outputs may also be connected back to their inputs to produce miniature, closed ecosystems—a principle followed (with mixed success) in Arizona's Biosphere and in NASA projects for interplanetary spacecraft.[10] Increasingly, the flows that they channel are monitored by sensors, precisely controlled by valves and switches, filtered and tempered in a multitude of ways, and managed by sophisticated digital systems. One way or another, the pipes form linkages between small-scale metabolic processes within my skin and larger-scale processes outside it.

My sexual plumbing is constructed to interface with other, compatible sexual plumbing for the efficient transfer of genetic information in fluid format. Unfortunately, the fleshware connection can be flaky, unstable, and nonstandard (worse than a dial-up modem), but there are numerous illustrated manuals describing recommended configurations and protocols. I am a node in a body-to-body network that, sadly, turns out to be effectively organized for virus propagation as well.[11] Traditional forms of sexual union are circuit-switched and synchronous, with all the intensity and risk that this entails, but refrigerated sperm banks now function as genetic code servers. In vitro fertilization is an asynchronous transaction—the organic equivalent of downloading email, and about as arousing. Blood donations, banks, and transfusions form a similar fluid interchange network; early attempts at transfusion involved synchronous artery-to-vein links, but if I make a blood donation today, I upload to a blood bank and some anonymous recipient later downloads.[12] From the perspective of our genes and viruses, our bodies and their in vitro extensions are just temporary nodes in an evolving propagation network.

In the extravehicular mobility units (EMUs) worn by the Apollo astronauts, internal and external plumbing systems were locked in a tight, semi-permanent, crypto-sexual embrace. NASA diagrams show backpack primary life support systems (PLSSs), with supply and removal systems, intimately plugged into the bodies of the moon walkers, and controlled from chest-mounted consoles.[13] The necessary interfaces were maintained by a maximum absorbtion garment (MAG) to collect urine, a liquid cooling and ventilation garment (LCVIG) to remove excess body heat, an EMU electrical harness (EEH) to provide communication and bioinstrument connections, a communications carrier assembly (CCA) for microphones and earphones, an in-suit drink bag (IDB), and a polycarbonate helmet with oxygen supply and carbon dioxide purge valve. If you had the right stuff, you not only walked on the moon, you got to sleep with your extrabiological body double—snugly beneath two layers of inner cooling garment, two layers of pressure garment, eight layers of thermal micrometeoroid protection garment, and the outer cover.

In everyday life, of course, the linkages are a bit looser. Unless I find myself on extracorporeal life support I am only intermittently plugged in.[14] But (in developed urban environments, at least) the interface points—water faucets, supply and return air registers, domestic refrigerators, baths, sinks, and showers, garbage disposal units, gasoline pumps, urinals, and flush toilets—are never far away. The preindustrial equivalents of these points, such as the well and outhouse, or the seats of ease on a sailing ship, were generally less pleasant and sophisticated, and were kept at a distance, but their modern descendants have moved indoors to become standard, indispensable organs of buildings.[15] The large-scale construction of these intestinal extranets and the integration of their interfaces into architecture were among the heroic projects of early modernism; they were conceived (though many are now more skeptical) as progressive enterprises bringing hygiene, equality, and cohesiveness to the industrial city.[16] Eventually, as *Ulysses* obsessively emphasizes, there was no difference between shitting on a Dublin cuckstool and defecating in a London toilet. And the sewers poisoned the oysters just the same.

By the 1960s, the architectural avant-garde had begun to take note of all this. It was sensing a shift from composition of space and structure—the conception of architecture that had been expressed by Laugier's conspicuously unserviced "primitive hut"—to the construction of these pipe, duct, wire, and mechanical movement networks. (That was, of course, where the money was now going in most buildings.) Archigram in Britain, Superstudio in Italy, François Dallegret in Canada, and others produced endless images of human bodies—mostly young, photogenic, and minimally though fashionably clothed—surrounded by elegant plumbing and ductwork, large-scale mechanical contraptions, and places to connect to systems that supposedly would supply whatever you wanted on demand.[17] Renzo Piano and Richard Rogers built the Pompidou Center in Paris, with its ductwork and mechanical systems boldly exposed, Reyner Banham engagingly supplied some theory, and the imagery eventually made it to the movies in *Brazil*. In retrospect, it is easy to see that this (mostly) cheerily optimistic brand of technofetishism got half the story right; networks were, indeed, increasingly crucial.[18] What it missed—dooming its brightly colored, hard-edged images of Capsule Homes, Plug-in Cities, Instant Cities, Cushicles, Suitaloons, Manzaks, Rokplugs and Logplugs to seem closer, now, to Jules Verne than to William Gibson or Neal Stephenson—was the emerging role of hyperminiaturization, wirelessness, digitization, and dematerialization.

SENSORIUM (AUGMENTED)

Telephones, as the remaining McLuhanistas keep assuring us, are interfaces to yet another network infrastructure—one that that now stretches my speech production and reception system around the globe and multiplies its points of presence. It didn't seem quite that way at first, since the earliest models were large, heavy devices attached to walls and sometimes enclosed within celebratory booths—electrically powered descendants of the huge, earlike "listening systems" that Athanasius Kircher proposed to install in seventeenth-century palace walls.[19] They were components of buildings, and they established place-to-place networks. You were never quite sure who would pick up at the other end, and the relationship to our bodies was neither continuous nor intimate.

But later versions of the telephone were smaller and lighter, and they plugged into modular outlets; you could walk around with them, and the coordinates of the connection points became fuzzier. Handsets were scaled and shaped to the human jawbone (from front teeth to socket near the ear) and resided on desktop cradles. Now cellphones fit in a pocket, they never leave us, and (in some cultures, at least) they are never switched off. They may even be wired into our clothing and equipped with earsets (scaled and shaped to the *interior* of the ear) for hands-free use. They are more part of our bodies than part of the architecture.[20]

In much the same ways, my retinal receptors have been multiplied megafold by CCD arrays embedded in digital cameras, scanners, VCRs, Webcams, and videoconferencing systems.[21] Some of these visual receptors are handheld, others are built into vehicles (from automobiles to imaging satellites), and yet others are installed within buildings. Some even operate through inconspicuous pinholes in walls. Some work independently, but increasingly many are hooked into the worldwide, digital storage and distribution network.

With these pervasive audio and video sensing systems, the lines dividing electronic conversation, accidental overhearing, deliberate electronic eavesdropping, and systematic surveillance are thin ones— more a matter of context and intention than of technology. As wireless bandwidth increases, as the video equivalents of cellphones emerge, and as sound and image capture points proliferate, the balance is inexorably shifting toward surveillance.[22]

I am becoming the focal point of a global personal Panopticon. It is not a spoked building radiating from my body (that is, a network of one-way sightlines), as constructed by Jeremy Bentham's Enlightenment imagination and elevated to iconic status by Michel Foucault, but a wildly ramifying circuit structure with artificial eyeballs at the ends of the wires.[23] There are even tiny, battery-operated wireless eyes that can be left anywhere and will transmit whatever they see to the nearest Internet reception point.[24] There are wireless video camera pills (about the size of a vitamin E capsule) that transmit images of the small intestine to a data recorder worn on a belt.[25] And as autonomous, nomadic eyes get even smaller, they will be mounted on remotely controlled micro-robots or insects (most likely cockroaches) with electronic implants.[26]

Although audio and video sensors are most evident to us in our daily lives, electronic sensing is by no means limited to the acoustic and visual domains. Air-conditioning systems depend upon temperature and humidity sensors. Vacuum cleaners and washing machines contain pressure sensors. Accelerometers, orientation detectors, inclination detectors, and vibration detectors can track motion. Strain gauges tell how a structure is behaving. Sensors for chemicals and biological agents provide the rough equivalents of our senses of taste and smell. In general, any self-contained device that detects a property and produces a signal is a sensor that I can connect to a network and use to extend my powers of observation and surveillance.

GAZE (UNRESTRICTED)

I am both a surveying subject at the center of my electronic web and the object of multimodal electronic surveillance. All of those constructions of the gaze that the post-Foucauldians have alerted us to— the gaze of desire, the gendered gaze, the consumer's gaze, the critical gaze, the reflexive gaze, and certainly the gaze of power—are extended, reorganized, and reconstructed electronically. Re-released Big Brother (or Big Other 2.0) is made from little pieces linked together; he/she is everywhere and all of us—at least when we pay attention. And combating the unwanted gaze or audience is no longer a matter of proximity and enclosure—of hushed voices, drawn veils, and retreating behind closed doors—but of controlling access to networks, databases, and messages.

Furthermore, the observer need no longer be an embodied subject hunkered homunculus-like within an enclosure, like Kircher's palace courtier, Bentham's jailer, a camera obscura peeper, or the Wizard of Oz. Nor is it necessarily a bunch of bored guys peering at flickering video screens in a security command center—the modernist imagination's icon of surveillance. It may be a dispersed observer swarm, as with cellphone-equipped, celebrity-spotting teenagers. Or, as in the U.S. National Security Agency's Echelon and Carnivore systems, the observing mechanism may be software that filters streams of audio, video, or text data to recognize and extract objects and events of interest.[27] (A stream containing the words "White House" and "attack" is likely to attract attention.) Furthermore, as digital records accumulate on the Web and in other types of online databases, they may be sorted, searched, fused, and filtered in numerous ways; surveillance may be conducted both in real time and asynchronously. These strategies overcome the Orwellian Big Brother's very human limitation—that he could only pay attention to a few things at a time.

Since the shrinkage of microphones, video cameras, and other sensing devices makes them increasingly invisible, since it is usually impossible to tell what they are connected to anyway (particularly when they are wireless), and since surveillance software seldom snoozes, our mere knowledge of the widespread existence of surveillance apparatus establishes the presumption of invisible, anonymous, unverifiable observation. Such surveillance is, as Foucault put it, "permanent in its effects, even if it is discontinuous in its action." We act as if we are observed, even when we may not actually be. That is harmless enough when invisible traffic cameras discourage me from speeding, and it can be reassuring if I know that I may need emergency assistance at any moment, but it is easy to imagine more sinister uses of this new, pervasive machinery of discipline and control. Although Foucault wrote before the digital revolution exploded, he knew in his bones what was coming; jails could morph from enclosures to networks, and in their characteristic postmodern form they could be larger and more totalizing than anything Bentham might have imagined.

Under these conditions, the traditional distinctions between a city's public and private spaces are eroding. When Nolli made his famous map of Rome, the difference seemed straightforward; he could show the citywide network of public spaces (streets, piazzas, and the interiors of churches) in white, with all their details, while shading private spaces in anonymous gray. In our secular age, new Nolli-style maps would, of course, depict the interiors of shops and malls rather than those of churches. They would also show zones of hypervisibility— public areas subject to video surveillance and maybe to the scrutiny of face recognition software that picks out putatively undesirable characters. Conversely, there would be zones of spatial aporia—the discreetly anonymous, secure sites of the servers and telecommunications hubs that make everything work.

I can also peer electronically from private spaces into public ones and from public spaces into otherwise private ones, creating scrambled and sometimes-paradoxical civic conditions. Security cameras provide interior private spaces with one-way views of public exteriors, while exterior displays occasionally reveal what's going on within a building—often for the benefit of those who cannot be accommodated inside. Two-way videoconferences can link public to public, connect private to private, or electronically mix public and private. New genres of electronic exhibitionism and voyeurism, such as dorm room Webcams and reality television shows, put private spaces on public display. The space of television broadcast studios is normally kept physically private (that is, closed to direct view and surrounded by strict security) while electronically presenting itself publicly. On the other hand, sports telecasts make stadiums recursively public; we can electronically watch audiences physically watching the game. Audio headsets can create private acoustic bubbles in the midst of public spaces, and video headsets can create even more dramatic disjunctions.

Wireless video cameras, which are designed to transmit to nearby base stations, blur Nolli maps in particularly insidious ways. If I am in the neighborhood, I can, if I wish, intercept these transmissions and display them on my wireless laptop computer. As I move around the city, I can surreptitiously open wireless windows at will.

The shift from naked eyeballs to networked video also changes the rules of public space layout. As the urban theorist Camillo Sitte observed, the winding streets and odd-shaped plazas of medieval cities continually provided visual surprises—not to mention opportunities for unexpected encounters and ambushes. But the streets of Haussmann's Paris and L'Enfant's Washington slice through the urban fabric in long, radiating, straight lines; rather than surprises, they provide continuous views of distant monuments, and instead of allowing for easy ambush, they are designed for efficient military surveillance, deployment, lines of fire, and control. They are, in other words, scaled-up panoptic diagrams. Now, though, these spatial moves are unnecessary; following September 11 and repeated terror alerts from the Justice Department, the Washington, D.C., Police Department began to install surveillance cameras in Metro stations, public schools, street intersections, shopping areas, and residential districts.[28] The resulting video feeds were viewable both at stationary command centers and in squad cars. With a bit more wireless infrastructure, they would, no doubt, go directly to cellphones. For the cops, camera locations suddenly became more important than street geometry, and the scope of their gaze was disconnected from urban geography. If you have electronic sightlines, you no longer need baroque street networks.

Steven Spielberg's 2002 film *Minority Report* vividly extrapolates this emerging condition.[29] In his Washington of 2052, transportation systems do not flow along L'Enfant's avenues, but employ flying machines that can swiftly go anywhere and automobiles that can drive up the sides of buildings. Electronic surveillance systems are installed in every doorway and are carried by tiny, spiderlike, heat-sensing, wireless-communicating robots that can squeeze through the cracks under doorways. Retina scans from these systems are electronically checked against databases to track the movements of citizens. And asynchronous surveillance is carried to a new level—precrime extends it to the future as well as the past.

Once, the natural condition of cities was opacity; architects created limited transparency by means of door and window openings, enfilades, open rooms, and public spaces. Today, the default condition is electronic transparency, and you have to work hard to produce limited zones of privacy.[30]

My augmented nervous system, like that of the D.C. police, has immeasurably transcended the disposition of my flesh.[31] It has extended itself electronically by means of copper wires, fiber-optic cables, and wireless channels that connect my brain to electronic memory, processing points, sensors, and actuators distributed throughout the world and even in outer space.[32] It has been expanded to sense—where necessary—not only visible light but also infrared, ultraviolet, and very low-intensity light, to make visible the tiniest of objects, to capture sounds far outside the audible spectrum, and (through MEMS technology) to be acutely sensitive to odors, vibrations, accelerations, pressure and temperature variations, and just about anything else that may be of interest or importance to me. The firewall of my skin is crossed by electronic and electromechanical interfaces to my hands, eyes, and ears—and occasionally to other organs as well. Some of these interfaces are permanently active, others are switched on and off as required. Some are deployed at fixed locations within my surrounding environment, some are portable, some are wearable, and some may take the form of miniaturized implants.

And it does not stop there; we are on an accelerating curve. Contemplating the recently invented telegraph, Nathaniel Hawthorne began to imagine a worldwide system of artificial nerves, endlessly pulsing with electrical impulses.[33] Views from the Apollo spacecraft taught us to see the entire globe—a sparkling blue ball in the dark void—as an object of comprehensive electronic surveillance. Today's commercial observation satellites, in the words of RAND analysts, "promise to bolster global transparency by offering unprecedented access to accurate and timely information on important developments,"[34] and their military equivalents are even more capable.

Sensors of all kinds are now becoming tiny, inexpensive, and network-enabled, and they are increasingly being integrated into very large-scale sensing systems. By 2001 a committee of the National Academy of Sciences could confidently suggest, "Networks comprising thousands or millions of sensors could monitor the environment, the battlefield, or the factory floor; smart spaces containing hundreds of smart surfaces and intelligent appliances could provide access to computational resources."[35] Oak Ridge National Laboratory was pitching Sensor Net, consisting of biological, radiological, and chemical weapons detectors mounted on cellphone towers.[36] Meanwhile, William Gibson's fictional heroes inhabit a world in which neural extension is taken to the limit by dispensing with all the intermediate junk and just jacking brains directly into the global network.

So my sensorium is no longer localized by the inexorable laws of visual occlusion and acoustic decay, the range of my exploring fingertips, and the wavelengths and scales to which evolution has tuned my original sensory equipment. It reaches to wherever there are sensors with network connections.[37] My experience of places and events depends decreasingly upon positioning my eyeballs at precisely chosen locations (as Renaissance perspective implicitly insists) and increasingly upon electronic access to a globally dispersed, multimodal sensing and reporting system. And, as this system continually gets denser, the relevant metaphor is no longer that of the all-seeing eye (as depicted, for example, on the U.S. dollar bill), but that of a continuous sensate skin. The earth itself is growing such a skin, the surfaces of buildings are beginning to evolve in that direction, and our clothing will eventually go the same way.[38]

The radical de-localization of our interactions with places, things, and one another—in space through electronic sensing and telecommunication and high-speed travel, and in time through electronic and other forms of storage—was identified by Anthony Giddens as one of the characterizing features of modernity.[39] If we lived within the walls of a preliterate, ancient city, all of our interactions would be face-to-face and synchronous, conducted in places like the agora; but we now live at the nodes of networks that allow a great many of our interactions to be remote and asynchronous. With the continued shift from enclosures to networks, we have bolted beyond modernity's spatial and temporal extensions to a condition of global hyperconnectivity.

CONTROL (DISTRIBUTED)

When I move a cursor with a mouse, I execute simple actions manually and observe the results of those actions directly in front of my nose; the feedback loop that keeps me in control, and allows me to learn from what I'm doing, operates at a scale of centimeters. When I use a television remote, I send commands electronically and get visual and audio feedback; I depend upon a direct line of sight and being within earshot, and the feedback loop operates at a scale of meters. It's the same when I operate my five-year-old's radio-controlled toy racecar; I'm in trouble if the vehicle disappears behind the couch and I lose the feedback loop. But it's different when I operate a video camera, telescope, or robot over the Internet; in this case, the feedback loop consists of bits flowing through a network, and it may be effective over thousands of kilometers.

In many contexts, form follows feedback. It is now a commonplace of control theory that swarms of bees, schools of fish, and flocks of birds are held together by short-range feedback loops. Slime mold cells follow pheromone gradients to aggregate, and to respond to global changes in their environment, at even smaller scale. These collectives are self-organizing; they don't need leaders.[40] Their spatial coherence and their complex, adaptive behaviors emerge from the capacities of simple agents to observe the movements of a few neighbors and to adjust their own movements accordingly. Swarms of SMS-equipped urban teenagers are not very different—except that the electronic feedback loops linking their actions extend beyond their line of sight, maybe for many kilometers.[41]

Kids who hang together by pinging their posses in this fashion may behave in coordinated, purposeful ways, but they only occasionally form compact, readily identifiable, directly observable physical groupings at particular locations. Most of the time, the groups are spatially dispersed and invisibly linked. It is harder (though often appropriate) to conceive of them as discrete "things" that somehow contain goals.

Markets have gone the same way. They were once enclosed within well-bounded physical places, such as market squares. Information about demands, prices, and wares on offer was visibly and audibly evident within those places, and circulated swiftly through word-of-mouth as well. The invisible hand operated at very close range. But when I purchase theater tickets through an online auction site, or trade stocks electronically, I participate in a spatially dispersed market that relies upon long-distance feedback loops carried mostly through electronic linkages. It is an extended, distributed computational mechanism, structured by complex rules and integrating both humans and machines.[42] It is certainly a real and important thing, but you cannot literally go to it or even point out its location; it does not have a stable, definite physical identity.

Games have dispersed even more dramatically. A tennis or basketball court is a rigorously standardized piece of geography designed to contain the rule-governed flow of action; an "out of bounds" condition is not to be taken lightly. The players observe the ball and one another directly, at close range, and they respond accordingly. When I participate in an online computer game, though, I don't have any idea where the physical boundaries are or how many active players there might currently be; I interact with the other players, avatar-to-avatar, in software-generated, software-ruled virtual terrain. (There may be trouble if the value of a variable inadvertently goes out of range, or if a stack overflows—but that is another story.)

In the online ur-game Spacewar (circa 1961), the virtual terrain was a simple diagram, players were few, and feedback was crude and slow. By the days of Doom (which suddenly began to clog networks at the end of 1993) and Quake, the virtual terrain was in full-color 3D, there were thousands of players, and you had to respond instantly to feedback in the form of fearsomely skinned, weapon-wielding, animated warriors. By the turn of the millennium we were seeing massively multiplayer, persistent online worlds (Everquest, Ultima Online, Asheron's Call, and more). These often grew to be larger than the physical cities inhabited by their players, and they could exhibit complex social dynamics—such as the spontaneous formation of combat clans. In the subgenre of online sports games, such as NFL 2K3, the action took place in careful simulations of famous arenas, using simulations of professional players as avatars; geographically scattered players could meet in online "lobbies" to form pickup teams. In more intellectually oriented online worlds, such as that of the Slashdot news and discussion community, the model was the seminar room rather than the gridiron, and there was less interest in visual simulation, but feedback mechanisms and self-organization played a comparably crucial role.[43]

It isn't that such complex, feedback-rich, self-organizing systems don't have identifiable units and hierarchies. In fact, they usually do. But their units and hierarchies *emerge* from the dynamics of interaction; they are not predefined, like those of armies and corporations.

When social scientists first embraced the concept of feedback, they often thought in terms of homeostasis—of closed social groups regulated by negative feedback loops and seeking stable states, as a thermostat-controlled heating system seeks a stable temperature: community as a blob of slime mold. But that is not the way it has worked out in the network era. With the electronic de-localization of my interactions, the feedback loops that guide my actions and teach me about the world may operate at all ranges—from millimeters to thousands of kilometers. They may be synchronous or asynchronous. They do not integrate me into a single, closed, stable community. Rather, they engage me (with varying degrees of commitment) with multiple, scattered, perhaps spatially indefinite, and maybe transitory social and economic structures.[44] And they implicate me with innumerable, simultaneous, spatially overlaid patterns of self-organization in complex, unstable geographies and choreographies of control.

MIND (MULTIPLIED)

My capacity for awareness, response, and agency within these structures is a variable. My local stock of neurons has (the neuroscientists gloomily assure me) been diminishing as I grow older, but the supply of silicon and software at my disposal has been growing rapidly. Consequently, the neural network inside my cranium outsources more and more mental functions. I don't do much mental arithmetic any more; calculators and computers take care of that. I don't rack my brain for half-remembered facts; I look them up on the Web. I routinely exist in the condition that J. C. R. Licklider presciently identified, way back in 1960, as "man-computer symbiosis"—except that Licklider, Doug Engelbart, Ivan Sutherland, and other pioneers of interactive computing mostly had dialogue with desktop workstations in mind, whereas I now interact with sensate, intelligent, interconnected devices scattered throughout my environment.[45] And increasingly I just don't think of this as computer interaction.

I don't directly control all the functions of the machines and devices I use; I rely on the intermediating machine intelligence embedded in my cellphone, my car, my domestic appliances, the operating system of my laptop computer, and my software agents. Mostly I cannot tell whether such intelligence is supplied by local devices, by remote servers, or by some combination of the two, and it doesn't matter—as long as there is capacity available somewhere, and the connections are sufficiently fast. Often I cannot even tell whether a verbal response I receive over the network has been generated by a person or by a machine; the Turing test has stealthily been aced.[46] And when I'm in a deathmatch on a first-person shooter like Quake, an AI opponent such as ReaperBot may be more formidable than a human one. As nodes of machine intelligence are distributed just about everywhere, as electronic interconnectivity grows, and as electronic feedback loops multiply, cities are evolving into extended minds and biological brains are becoming elements of larger cognitive systems. It is Santa Cruz guru Gregory Bateson's "ecology of mind," but with much more silicon and many more electronic interconnections than he ever imagined.

Mary Catherine Bateson, looking back upon her father's intellectual legacy, has summarized his brilliantly insightful (if sometimes frustratingly fuzzy) formulations of this idea as follows: "A mind can include nonliving elements as well as multiple organisms, may function for brief as well as extended periods, is not necessarily defined by a boundary such as an envelope of skin, and consciousness, if present at all, is always only partial."[47] Bateson himself put it: "Within Mind in the widest sense there will be a hierarchy of subsystems, any one of which we can call an individual mind."[48]

He went on to ask, "But what about 'me'"? And his answer was uncompromising:

> Suppose I am a blind man, and I use a stick. I go tap, tap, tap. Where do I start? Is my mental system bounded by the handle of the stick? Is it bounded by my skin? Does it start halfway up the stick? Does it start at the tip of the stick? But these are nonsense questions. The stick is a pathway along which transformations of difference are being transmitted. The way to delineate the system is to draw the limiting

line in such a way that you do not cut any of these pathways in ways which leave things inexplicable. If what you are trying to explain is a given piece of behavior, such as the locomotion of the blind man, then, for this purpose, you will need the street, the stick, the man; the street, the stick, and so on, round and round.[49]

Similar spatial and functional extensions of mental systems are produced, in variously specialized ways, by the artist's pencil, the surgeon's scalpel, the diner's cutlery, the canoeist's oar, the fencer's sword, the tennis player's racquet, the boxer's glove, and the mother's Q-tip in a baby's ear. With surfboards and skateboards, feet instead of hands are the crucial connection points. Cyclists, sailors, and hang glider pilots interact with their environments at more points and through more complex mechanical linkages. Automobile drivers and airplane pilots do so through powered mechanisms, and in more advanced vehicles through electronic connections, interface devices, sensors, and servos. For several decades now, computer network users have done so through indefinitely extended electronic systems with distributed intelligence, elaborate sensor and effector systems, and increasing capacity for autonomous agency. Today, through miniaturization and wireless interconnection, these networks are becoming denser and more pervasive. In all of these cases the extended body functions as a system for perceiving one's environment in particular ways, reasoning about it, and executing actions that engage it.

It is not that there is a unified "world brain" of the sort rather crudely imagined by H. G. Wells in the 1930s.[50] But skulls and skins do not bound mental systems, and through computer networking these systems can now extend indefinitely.

MEMORY (EVOLVING)

When such extended and distributed mental systems have the capacity to store and recall information, they don't just live in the present. They can learn from experience and grow smarter over time. Specialized memory organs may, of course, be enclosed within craniums, but they may also be inscribed on stone tablets or sheets of paper or implemented electronically in silicon or magnetic oxide.

Extracranial memories once consisted of records clustered within easy reach, like the papers on my desktop or the notes and pictures on my refrigerator door, but they are evolving into global, self-organizing networks. One part of my artificially extended memory consists of the world's scholarly and scientific literature—a collection of paper documents, most of which exist in multiple copies and locations, cross-connected by citations. I access information in this form by physically tracking down particular pieces of paper. Another part consists of the World Wide Web, that is, of hyperlinked online documents that I access either by following links or by employing the indexes constructed by search engines. The two parts are increasingly interwoven; paper documents like this book reference Web pages, Web pages cite paper documents, and some documents exist simultaneously in paper and electronic form. Every day, millions of network nodes suck information into the growing online component of global memory, millions of users construct links, and as this whole, complex, hybrid network evolves, Brewster Kahle's Internet Archive grabs and stores snapshots.[51]

The larger this exponentially expanding, continually restructuring, networked, shared collector and memory becomes, the less escapes it.[52] It can respond to major events instantly, automatically, and on a massive scale. When the attacks on New York and Washington took place on September 11, 2001, related bits immediately began to flow in—not just from the immediate vicinities, but from around the world. There was an explosive accumulation of eyewitness footage and accounts, news reports and analyses, memorial sites, tribute pages, survivor registries, commentaries, and policy analyses. A year later, when the first books on the attacks were appearing, and discussions of an architectural memorial were just getting under way, the Internet Archive already had more than five terabytes of online, indexed data on the subject.[53] (For comparison, the entire Library of Congress contains about twenty million books, or twenty terabytes of textual information.)

So we can extend mental systems by putting memory and processor chips in things, networking them, linking them to sensors and actuators, and constructing feedback loops, but it isn't *just* a matter of that. If we take Bateson's view, there is no clear distinction between internal cognitive processes and external computational ones. Furthermore, as physicists are increasingly insisting, there is also no straightforward distinction between computational processes and physical processes in general; you can think of *any* physical process as the computation of some function. (Set things up in the right way, and the whole apparatus of bits, circuits, and symbolic computation might eventually turn out to be a cumbersome and unnecessary intermediary. It may just be the wrong level of abstraction for harvesting computational capacity on a really large scale.) At the speculative edge of this discourse, John Archibald Wheeler teases us to imagine "it from bit," Stephen Wolfram conceives of unfolding reality as the enaction of an algorithm,[54] and Seth Lloyd estimates the number of logical operations performed by the universe since the Big Bang.[55] If every particle is an automaton communicating with its neighbors, physical reality at the subatomic scale is just one, inconceivably vast network.

Opponents of artificial intelligence and skeptics about telepresence frequently argue that, without a body, an information processing system can learn and understand very little about its environment.[56] They have a point; the isolated "giant brains" of the midcentury modernist imagination are insufficient. Hal doesn't hack it. But the necessary embodiment isn't just that of the naked newborn infant—though babies are on the right track. And it isn't just that of Adam, losing his innocence in a God-given garden. We perceive, act, learn, and know through the mechanically, electronically, and otherwise extended bodies and memories that we construct and reconstruct for ourselves. And, as we are beginning to see, there is no clear limit to this extension.

INDIVIDUALS (INDEFINITE)

As Bateson had begun to realize, we are not fully contained within our skins; our extended networks and fragmented habitats make us spatially and temporally indefinite entities. His central insight was that the ancient distinctions between user and tool, building and inhabitant, or city and citizen, no longer serve us well. And at a practical design level, his point becomes ever more urgent as carbon/silicon interfaces get tighter, as networked silicon intelligence embeds itself everywhere, as EmNets (networked systems of embedded computers)[57] replace stand-alone boxes of electronics, as different types of networks are integrated into multifunctional systems, and as computer and cognitive scientists increasingly theorize "societies of mind" rather than discrete, unified intelligences.[58] We will do better to take the unit of subjectivity, and of survival, to be the biological individual *plus* its extensions and interconnections.

So I am not Vitruvian man, enclosed within a single perfect circle, looking out at the world from my personal perspective coordinates and, simultaneously, providing the measure of all things. Nor am I, as architectural phenomenologists would have it, an autonomous, self-sufficient, biologically embodied subject encountering, objectifying, and responding to my immediate environment. I construct, and I am constructed, in a mutually recursive process that continually engages my fluid, permeable boundaries and my endlessly ramifying networks. I am a spatially extended cyborg.

3

WIRELESS BIPEDS

As my networks extend ever outward, my circumscriptions multiply and expand like ripples in a pond. My cyborg self is structured— Linux-like—as a system of nested shells, with carefully articulated and controlled interconnections among the levels.

Under my epidermis there is a tightly packed, carbon-based kernel, mostly run by genetic code and my central nervous system but maybe augmented by implants. Then there is a wearable layer of cloth, leather, plastic, some metal, and a growing number of tiny machines and miniature electronic devices; to function as a coordinated system, these components need networking through my body or my clothing. When I travel in a vehicle, there is an additional, mostly metal layer, with its own increasingly sophisticated electronics, code structures, and interconnectivity. The architectural layers (which, of course, were Le Corbusier's idea of a *machine à habiter*) are generally composed of heavy materials, together with fixed-in-place machines and pipe, duct, and wire networks. Finally, the regional and global layers are formed by large-scale, long-distance infrastructure and geographically dispersed networks.

MIGRATION OF FUNCTIONS

Functions, and the artificial organs that provide them, may migrate back and forth among these layers as circumstances require. Consider protection and climate control, for example. If I need to keep warm, I can put on a sweater and rely upon body heat, or I can insulate the walls of my house and introduce a mechanical heating system. In the 1960s, Reyner Banham and François Dallegret dramatized this point in a famous project; their drawing showed them sitting, naked, in a transparent, plastic bubble that was inflated by the air-conditioning apparatus in the center. Buckminster Fuller went further and proposed a giant, climate-controlled, transparent dome over all of Manhattan. Marshall McLuhan hubristically trumped even that by suggesting that "global thermostats could by-pass those extensions of skin and body we call houses."[1]

For protection from the rain, I can use a portable umbrella or a permanent roof. If I worry about getting shot, I can bullet-proof my vest or the windows of my car, or I can post guards to take away guns at the door. When I ride in an automobile I depend upon the airbag in the dashboard, but that does not work on a motorcycle, so the airbag goes in my protective clothing. If I participate in a street protest, I can build a barricade or march behind a mobile shield. If I am an astronaut, I can have my life-support systems in a close-fitting space suit or in the infrastructure of a space station. For protection against SARS, I can wear a face mask, try to live in a clean room, or attempt to throw a quarantine cordon around my city.

Control points can also shift. In the days of simple mechanical controls, such as doorknobs and rack-and-pinion steering wheels, there was direct linkage of controlling device to controlled mechanism, and this severely limited spatial freedom. But the telegraph introduced the possibility of remote control; the motion of the transmission key induced corresponding motion in the reception device. Today, if I want to actuate some device I might find the switch on the wall, on the device itself (as with desk lamps and television sets), or on a hand-held remote. And, as increasingly many devices get network connections and IP addresses, anything can potentially be controlled from just about anywhere.

Less obviously, storage and processing nodes may migrate as well. My electric power supply may come from a button-sized battery or a tiny generator in my shoe, from a larger battery or generator in my automobile, from photovoltaic or fuel cell generators in the building I occupy, or from an electric grid extending over a wide geographic region and incorporating many large power plants. The light I require to find my way may come from a ceiling fixture or a miner's helmet. If I am a scuba diver or spaceman, my oxygen is stored locally and circulated through wearable plumbing; if I am a motorist or airplane passenger, I rely upon an onboard air conditioning system; and if I inhabit a building, I depend upon natural air circulation from openings, upon a room air conditioner, or upon a central, ducted, air handling system.

For my information supply and processing power, I may access digital memory and processing capability in a wearable or portable device, I may get what I need from a personal computer or local network server, or I may download from a distant server to a "thin client."[2] I may fetch it directly from the supply point, or I may cache it at various intermediate storage locations for faster access.

Even sensing functions may shift. The temperature in a room I inhabit might be regulated by a sensor on the wall or by one on my body. If I want a health-monitoring system to know if I fall down, I might put motion detectors in the walls or accelerometers on my body. In early attempts to create intelligent environments, multiple sensors were connected to a single, central computer that processed inputs and responded; now it seems more effective to distribute processing power to the sensors themselves and to perform low-level interpretation tasks locally.

There is (as with many migrations) a strong sociopolitical dimension to all this. If you control temperature by adding and removing clothing layers, then you privilege autonomous individual choice, but if you go for Banham's plastic bubble, then you construct a situation in which the naked inhabitants have to negotiate with one another about the temperature setting. And Fuller probably did not consider the fact that the weather inside his Manhattan dome would become a matter of New York municipal politics.

SCALE AND RANGE

Scale matters. The independently moving subsystems that embed our biological bodies come in various sizes, and these sizes determine the potential locations of storage and processing sites—in inner, intermediate, or outer layers. Somewhere, there is a distinction between mobile subnetworks associated with our bodies and surrounding, fixed-in-place infrastructure networks.

The nomadic aborigines of the Australian desert represent one extreme.[3] For them, almost everything is in the peripheral layers—the natural infrastructure. They carry very little and wear almost nothing; this lightens and liberates the body, but it demands an extraordinary knowledge of the habitat, and a highly developed capacity to locate and exploit the water, food, shelter, and other resources that it offers. Nineteenth-century European explorers of the desert—equipped with camel-loads of supplies—did not have this capacity and frequently perished.[4]

Backpackers carry more than nomadic aborigines, but they still strictly limit themselves. They carefully calibrate their onboard storage capacity to provide themselves with sufficient independent means for survival between reconnection points. The bigger their pack, the greater their range, but they pay the penalty of increased effort and decreased agility.[5] Horses with saddlebags provide greater range again, but have significantly diminished flexibility; there are more places they cannot go. Wheelchairs for the sick and elderly offer more limited freedom of movement, but can provide room for sophisticated mechanisms and electronics, medical supplies, and even intravenous drips.

Mechanical transportation, providing the capacity to move larger packages of stuff around, shifts the balance still further. Compared to a horse, an automobile can accommodate many more storage and processing capabilities onboard, in inner layers, to create a larger and heavier zone of independence from surrounding resources and conditions. This is carried to extremes in SUVs with cupholders, limousines with bars, Winnebagos with kitchens and bathrooms, and dustbowl immigrants making for California in their overloaded jalopies. A modern jetliner is a flying restaurant and theater. A manned spacecraft, at roughly the scale of a house, carries everything it needs through the most hostile territory imaginable. An aircraft carrier is a slow-moving, independent system at the scale and complexity of a city. Ron Herron, in the 1960s, gleefully crafted *épater*-the-old-guys images of walking cities.

At the extreme are self-contained vehicles—from Noah's ark to the starship *Enterprise*—that must operate, for long periods, where there is *no* infrastructure to rely upon. The ships of seafaring explorers, such as James Cook, faced this condition. Cook's tiny vessel had to carry provisions for months at sea, along with antidotes for scurvy. Modern submarines are designed to operate for similar periods away from their bases. And spacecraft for Mars missions will need long-term, self-contained, recycling life support systems—designed in a particularly careful tradeoff of bulk and weight against capability and range.

LOCATIONAL OPTIONS

Given a particular scale and granularity of mobile cyborg systems, design decisions about where to locate storage and processing sites—in inner or outer layers, in the fixed infrastructure networks or the freely moving subnetworks—respond to many factors. Designers must consider intensity, urgency, and predictability of need, bulk and weight of storage containers, network speed and capacity, continuity or discontinuity of network connection, frequency of reconnection and resupply points, quantity and perishability of the resource being delivered, security requirements, and ease of duplication. There may be a need to achieve aggregation benefits and scale economies by locating sites where there is plenty of room. And there may be a political agenda—a desire to enhance individual autonomy by locating resources and functions in inner, personally managed layers, or conversely, to strengthen community by shifting them to outer, collectively managed layers.

Water is heavy and bulky, for example, and we usually don't need it instantly, so (fashionable young joggers and gym rats aside) we only carry around water bottles when there is no other option. Instead, we organize extensive systems of reservoirs and tanks, and deliver to points distributed densely throughout buildings and urban environments.[6] In many cultures, the activity of constructing, managing, and sustaining these systems has been a crucial focus of political activity and critical to social cohesion.

Electric power, on the other hand, is now required continuously to drive wearable and portable electronic devices, so many of us find ourselves carrying several miniaturized batteries. That's fine if we just need milliwatts or watts for a low-powered electronic device, but if we need kilowatts or megawatts for some big machine, we're probably stuck with a massive battery, a bulky generator, or a wire.

Stored digital information is different again; ease of duplication makes it convenient to keep portable local copies of files when network speed is limited. But when it is of a variety that quickly becomes out of date, and when it can be transferred almost instantaneously through high-bandwidth channels, it makes more sense to maintain it on central servers and deliver it where needed on demand. Or, as content delivery systems like that operated by Akamai have demonstrated, it may be efficient to duplicate and store content at numerous cache servers close to end users and use sophisticated optimization software to figure out the quickest delivery paths as requests are received from particular locations.[7]

If things you need frequently are too cumbersome to carry around, then you can sometimes rely upon grids of fixed access points. Thus public telephone boxes and booths were prominent features of many cities in pre-cellphone days; you could find one quickly when you needed to make a call. Today, we increasingly rely upon private, portable handsets, and these public access points have greatly diminished in importance. Conversely, many students on college campuses would rather check their email at public access points than carry relatively bulky wireless laptop computers to do so. And toilet training accomplishes a liberating shift from continual reliance upon wearable diapers to occasional use of the fixed infrastructure.

In general, if you make bigger moving boxes, you can carry around more stuff and mobilize more functions; you can pack more functionality into an SUV than a little two-seater, and more still into a mobile home. But the overall effect of recent technological development has been to shift the dividing line between highly functional stationary boxes (architecture) and less functional movable boxes— that is, vehicles, portable devices, wearables, and implants.[8] Miniaturization (particularly of electronic devices) allows designers to jam more functionality into small packages, and extensive networking reduces the distances between replenishment points. Weapons designers were among the first to realize this; many of the early efforts to miniaturize electronics were driven by the desire to replace human pilots with missile guidance systems for the long-range delivery of destruction. In the early 1970s, Archigram's David Greene and Mike Barnard saw where miniaturization was leading more generally; they imagined the "electric aborigine," speculated about "the possible influence of miniaturized electric hardware on lifestyles," and proclaimed— a bit ahead of their time—that "people are walking architecture."[9]

This effect is most dramatically evident with artificial hearts and other organs. Once bulky bedside devices that kept users tightly attached to the nonwalking architecture at fixed locations, they migrated to backpacks and strap-on harnesses, and have now, in some cases, become small enough to implant. Step by step, the bodies of formerly tethered, immobilized patients have been liberated.

THE MISSING LINK

Until recently, however, there was a critical missing link. Interconnections across network layers worked reasonably well from the global scale down to the walls, but then there was a gap. Our bodies temporarily lost connection to larger networks when they got up and moved around.

Interim storage devices—water bottles, rechargeable batteries, and chamber pots—can expand the body's range from a fixed network connection point. (This only works within constraints of carrying capacity and expiration date.) So can flexible pipes and wires. A garden hose keeps you connected to the water supply system as you walk around your backyard, and a power cord keeps you connected to the electricity grid as you vacuum the floor. More dramatically, deep-sea divers depend upon their air hoses. But you cannot remain tethered to an outlet as you walk city streets, drive an automobile, or pilot an airplane. And you cannot link satellites to ground stations with wires. In these contexts you must rely upon wireless connections.

The possibility of continuous wireless linkage first emerged in the mid-nineteenth century, when James Clerk Maxwell theorized the existence of electromagnetic disturbances that might serve such a purpose.[10] By 1888 Heinrich Hertz had experimentally employed sparks to produce radio waves, and by the early 1890s William Crookes—in a classic piece of futurology—could speculate in the *Fortnightly Review* about a world of wireless intercommunication.[11] At the turn of the century, through the pioneering work of Guglielmo Marconi and others, the telegraph finally became the wireless telegraph. Ships at sea were soon maintaining continuous telegraphic contact with shore stations, and by the 1920s police cars and taxis were getting primitive radio telephones for voice communication.

In the mid-1940s, Bell Labs developed the idea of distributing low-powered transmitters over wide areas and handing off calls among them to provide continuous mobile service to large numbers of users. This was the beginning of the cellular telephone, but commercial cellular systems were not deployed until the late 1970s. Since then, the growth of cellular systems has been explosive; by 1990 there were about eleven million cellular users worldwide, by 1995 the number of new mobile phone subscribers each year was beginning to exceed the number of new fixed telephone subscribers, and by the early 2000s the number of users was heading into the billions.[12]

Early cellular systems employed analog signals and were intended primarily to handle voice communications. By the early 2000s, systems were going digital and increasingly shifting their emphasis from providing continuous, person-to-person voice connections to handling bursts of chip-to-chip data. This generated interest in broadband wireless systems that could deliver data not just at rates of kilobits per second, but at megabits, or even hundreds of megabits per second—as required, for example, by sophisticated multimedia.[13] Eventually, the differences between voice and data wireless systems began, messily and haltingly, to disappear.[14] We entered a world of GSM and G3 cellphone service, IEEE 802.11a and 802.11b[15] local-area networks (the "wireless Internet"), Bluetooth[16] networks to replace the serial and USB cables that had interconnected adjacent electronic devices, and high-speed UWB[17] networks. In research laboratories, many other alternatives were being explored. There was no single wireless network, and there were competing and conflicting standards, but seamless, global, wireless interconnectivity now seemed within our grasp.

THE LOGIC OF WIRELESS COVERAGE

The goal of a wireless transmitter is to provide satisfactory reception to suitable devices (either fixed or mobile) within some target area. This is essentially a matter of transmission antenna type and placement, signal strength, signal frequency, receiver design, the need to make efficient use of available spectrum, the requirement for sufficient spatial or temporal separation of signals that occupy the same parts of the spectrum, and policy governing spectrum use.[18] It turns out that there is a complex logic of wireless coverage, and that this logic has motivated the construction of wireless infrastructure "shells" encircling the earth at successively greater altitudes, like the layers of a rather messy and incomplete onion. The coverage areas of wireless systems overlap with one another, and with domains established by geographic and political boundaries, to strengthen some established social and political groupings while subverting and weakening others.

The smallest, lowest-powered, shortest-range systems communicate over distances of centimeters or meters. There is little need to regulate their use of spectrum, since they are not likely to create interference with other systems. They obviously aren't useful for person-to-person voice communication (your unaided voice probably carries further), but they provide a convenient, flexible way to interconnect tiny computing devices without physical networking—an idea that has been explored in the MIT Media Laboratory's "pushpin" and "paintable" computer projects. And they can be organized to provide multihop connections over longer distances—much as packets are routed from node to node through the Internet.

At a slightly larger scale, Bluetooth-enabled devices (which contain special microchips) can interconnect wirelessly on desktops or within rooms—over distances up to about ten meters. Bluetooth was originally developed in the mid-1990s to link laptop computers to mobile phones. But it also now serves to interconnect numerous other types of consumer devices, such as MP3 players, digital cameras, printers, and video projectors, and to link portable devices to stationary network access points. Most interestingly, a Bluetooth device can establish instant connection to any other Bluetooth device that comes within range—thus enabling, for example, ad hoc networking among laptops in a conference room or classroom. Essentially, Bluetooth and similar systems provide freedom of movement and flexibility of spatial arrangement within room-sized socio-technical systems. Bluetooth is not the only wireless technology suitable for this role, and it is likely that better alternatives will supplant it, but it has at least served to introduce short-range wireless connectivity.

At the next scale up, Wi-Fi (802.11) and similar wireless base stations typically have ranges of 100 meters or so, but their signals are easily blocked by walls and other obstructions.[19] Thus they are well suited to providing mobile coverage within private homes, businesses, cafés, gardens and parks, and so on. As these systems are deployed, our built environment is being populated, increasingly densely, with wireless points of presence that link low-powered, miniaturized, portable devices to high-speed, long-distance networks. Base stations and portable devices provide personal mobility, while wiring in the walls provides indefinitely expandable capacity.[20]

Since these base station systems provide coverage at room and building scale, they tend to strengthen interconnections within established occupant communities. If they are open to visitors, they also extend a new form of hospitality. By providing free Internet access, they can add a new dimension to public space; Midtown Manhattan's Bryant Park was one of the first public places to offer this amenity. And since the technology is relatively inexpensive, simple to install, and uses unlicensed spectrum (in most places, anyway), it encourages grassroots neighborhood networking.[21] With more powerful base stations, it is also well suited for municipal efforts by small towns and villages, electronically updating the communal role of the central church bell tower or the minaret; in 2002 the small Georgia town of Ellaville (population 1,700) pioneered this strategy by installing a broadband base station on its water tower, together with rooftop antennas, to provide Internet connectivity within a kilometer radius. The New Zealand city of Auckland deployed a larger-scale, demonstration network throughout its central business district. However, there is a price to pay for this electronic conviviality; since signals do not stop *precisely* at walls, there are also dangers of overlap and interference between coverage areas, of unauthorized appropriation of base station capacity by neighbors or passersby, and of electronic eavesdropping.

Outdoors, at city-block and urban neighborhood scales, cellular infrastructure, which makes use of licensed spectrum and centralized switching, typically begins to take over. The infrastructure (particularly for advanced digital service) is costly, so it is typically deployed and controlled by telecommunications companies rather than by grassroots groups. Each base station in the system comprises a transmitter, a receiver, and a control unit, and is located roughly at the center of a cell. Coverage areas consist of mosaics of such cells. Cells may be about ten kilometers across, but they will be subdivided into smaller cells (with correspondingly lower-powered transmitters) in areas of high usage. The base stations for the smallest microcells are found on lamp posts and other street furniture, those for larger cells are often mounted on small buildings and special towers, while those for the largest cells move to the tops of hills and skyscrapers. Since suitable base station locations are in limited supply, and since there are often competing cellular systems, there tends to be increasingly intense contention for cellular infrastructure sites.

Although the infrastructure of cellular grids has traditionally been deployed at fixed locations, mobile "cells on wheels" may be used for rapid disaster recovery (these were much in evidence in Lower Manhattan in the days following the World Trade Center attacks), and there has been growing interest in the possibility of ad hoc, towerless cellular networks that are carried by mobile handsets—and thus automatically follow users around and adjust for density.

Naturally enough, cellular providers have tended to concentrate their infrastructure in high-density urban areas, and along highly traveled transportation routes, where the payoff is greatest. In the developing world, and in sparsely populated areas, this has produced a pattern of urban wireless "islands" connected by long-distance links. Occasionally, as with GrameenPhone's GSM cellular system in Bangladesh, an explicit commitment to the welfare of the rural poor has generated more even coverage.[22]

The next scale of wireless infrastructure is that of high-powered, tower-based, licensed transmitters providing coverage over distances of tens, hundreds, or even thousands of kilometers. This type of infrastructure began to emerge very early, with the first wireless telegraph towers—which were thought of, for a while, as "electromagnetic lighthouses." These were followed by the towers of early mobile radio telephone systems, such as those used by taxis and police cars. And chains of microwave transmitter/receiver towers have sometimes been employed (particularly in difficult terrain) as an alternative to long-distance telecommunication cables.

Socially and politically, though, a more interesting use of high-powered, tower-based infrastructure is for radio and television broadcasting to urban, regional, national, and even global audiences. Where the industrial-era combination of machine-powered presses with rapid transportation had created urban mass audiences that were reachable within hours, this type of infrastructure provides synchronous access to such audiences at very low cost. Since electromagnetic spectrum is a finite resource, there can only be a limited number of broadcasters in a given geographic area without interference, so this type of infrastructure tends to hegemony—concentrating political power and cultural influence in the hands of those who control the towers.[23] Consequently, governments have generally sought either to retain direct control of the transmission towers themselves or to license control to a few broadcasters.[24] In general, the taller the transmission tower you control, and the more powerful its signal, the wider your broadcast coverage; thus the prominent towers that rise proudly from the tops of the tallest skyscrapers in cities like New York are not only instruments of broadcast dominance, they are visible announcements of it.

When a high-powered system is used for two-way communications within some group that is scattered across a wide geographic area, different speakers take turns occupying the limited number of available channels (often just a single channel). Shared, structured use of a common resource in this way provides cohesion to communities as diverse as those of the shortwave School of the Air in the Australian outback, truckers chatting on their CB radios, taxi drivers and patrolling police officers going about their business, and Afghan fighters exchanging taunts via their walkie-talkies. The commonly available spectrum acts as a social focus, much like the well in a traditional village.

When Sputnik began to transmit signals back to earth in 1957, the possibility emerged of extremely high-altitude transmission points, with correspondingly large "footprints." By 1960 NASA had launched Echo 1—a Mylar balloon satellite that could reflect signals back to earth. (This was a variant on the older idea of bouncing signals off the moon.) In July 1962, Telstar, the first active communication satellite, went aloft and carried the first live, transatlantic television transmission. Passage of the Communications Satellite Act by the U.S. Congress quickly followed, and COMSAT (Communications Satellite Corporation) was formed. Not coincidentally, Marshall McLuhan's phrase "the global village" began to resonate with the popular imagination. In the succeeding decades, the skies were gradually blanketed with satellites.

Not far beyond the Earth's ionosphere, in orbit at distances between 500 and 2,000 kilometers, there are now low-earth-orbit (LEO) telecommunication satellite systems. Satellites in a LEO system contain transponders that receive uplinks from transmitters on the ground, converting them into downlinks to receivers on the ground or handing them off to other satellites. (A transponder is a device that receives a signal and transmits some sort of signal back.) The Iridium and Globalstar LEO systems were first put in place with great fanfare (maybe a bit prematurely) in the 1990s to extend the idea of a cellular network to the skies and to provide wireless voice and data systems with global coverage. So-called Little LEO systems, such as Orbcomm, are designed for paging, tracking, and similar applications of small bursts of data. Big LEO systems, such as Globalstar, work at higher frequencies, support higher data rates, and can support voice and positioning services.

In order to provide service at desired locations on the earth's surface, LEO systems must extend coverage to the *entire* earth's surface. Thus they are particularly important to isolated and sparsely populated areas that are not readily served by other types of infrastructure. Furthermore, there is likely to be considerable excess capacity in these areas—capacity that might, at least in principle, be devoted to education and to supporting economic development.

Further out, orbiting at 5,000 to 12,000 kilometers, are medium-earth-orbit (MEO) satellite systems, such as ICO. These operate much like LEO systems, but find a different balance of technical tradeoffs. They require fewer handoffs but higher-powered signals, and they introduce longer delays due to the greater distances involved.

Furthest out, at about 35,000 kilometers, are the geostationary telecommunications satellites. Unlike LEO and MEO satellites, they are stationary relative to the earth—much like *very* tall transmission towers. They provide footprints covering very large areas of the earth's surface, which is very effective for television broadcasting but wasteful of spectrum (and therefore expensive) for point-to-point communications. They also introduce delays that are very noticeable in synchronous voice and video communication. Proposed by Arthur C. Clarke in 1945, the first of them (Intelsat 1) was launched in April 1965, and they have proliferated enormously since. They now cluster densely over the more populated parts of the earth, providing services such as voice communication, digital video (for example, via the DBS system), and Internet access through services such as DirectPC and Starband.

Obviously there is competition among the various different types of satellite systems, and between satellite systems and terrestrial wireless systems.[25] When a GEO satellite is in position, it provides instant coverage over a very wide area, with significant performance and capacity limitations. LEO satellite systems have very high initial costs, and take more time to deploy, but promise technical advantages. Terrestrial systems are relatively inexpensive and can be extended incrementally; if this happens rapidly while new satellite systems are being planned and deployed (as the planners of Iridium discovered, to their cost), much of the potential market for satellite service is lost. In the long run, satellite systems seem likely to occupy some important niches in wireless service (such as GPS, global paging, and service to sparsely populated rural areas), but not to provide a universal solution.

From centimeter-range micro-wirelesses to broad-coverage geostationary satellites, the wireless world is weaving around itself an increasingly dense, multilayered cocoon of antennas, network access points, relay points, and channels. Different types of physical channels are increasingly being integrated, through telecommunication standards and protocols, into vast, seamless systems of bewildering complexity. Every point on the surface of the earth is now part of the Hertzian landscape—the product of innumerable transmissions and of the reflections and obstructions of those transmissions. The electromagnetic terrain that we have constructed, and continue to elaborate, consists of hotspots and deadspots, exposed areas and shielded areas, cells that get you through and overloaded cells that don't, signals (encoded in many different ways) that interfere with one another and signals that are cleverly multiplexed so that they don't interfere, jammed zones and Faraday cages, and the endless buzzes and bursts of electromagnetic noise.[26] It is an intricate, invisible landscape— one that is hinted at by the presence of antennas (sometimes, as well, by symbols warchalked in the street to indicate hotspots),[27] and can be made manifest by wardriving or warstrolling with a wireless laptop.

This landscape frames a complex geopolitics and political economy of wireless coverage. Within it, at every scale, there is competition for access to communities, for antenna sites, for timeslots, and for channel capacity.[28] Just as the kingdoms and empires of old struggled for control of terrestrial territory, those who seek power today increasingly contend for control of the airwaves.

One geopolitical strategy, with its roots in earlier traditions of telephone and broadcast communications, is to treat spectrum like frontier land. Governments chop it up and sell it—as in the spectrum auctions in many parts of the world that preceded introduction of G3 cellphone service. This facilitates comprehensive, top-down planning. However, it concentrates responsibility for providing coverage in the hands of a few license holders, and it encourages the development of centralized networks in which everything must flow through a few major switching centers, which become increasingly overloaded as the numbers of users grow. And it often slows down the extension of service, since spectrum auctions saddle license holders with huge debts.

A competing strategy, which draws upon the lessons of the Internet, is to think of spectrum as a communal resource, like the old village commons, or the land available to a squatter community. Anyone can use it, as long as they follow a few rules. This depends upon availability of unlicensed spectrum, and use of short-range, maybe multihop wireless technologies. It is messier, makes provision of adequate security harder, and risks overexploitation by a few at the expense of the many, but it has some important attractions. It allows the decentralized, bottom-up construction and interconnection of networks—much as packet switching and TCP/IP enabled the explosive, bottom-up formation of a global wired network. And it encourages a redundantly linked, decentralized network structure—which means that the addition of channels can enhance capacity without overloading central nodes.

There is, however, an additional twist. Since wireless spectrum is an immaterial, electronically managed resource, it can potentially (unlike land) be reallocated swiftly and automatically to meet changing demands. This opens up the possibility of dense wireless networks in which nodes cooperate dynamically to use available spectrum with maximum efficiency.[29] It seems likely that this will be the key to future expansion of wireless networks in densely populated areas.

At the extreme, deployment of wireless infrastructure can become a runaway, viral process. With very inexpensive wireless nodes, and a common standard such as 802.11, individuals can add nodes at will. With multihop technology, mobile and ad hoc wireless nodes can spontaneously form chains back to fixed infrastructure. And positive network externalities, supported by appropriate network architectures, can vigorously drive the expansion process; every node added to a network increases the value of existing nodes.

There is a close parallel, in all this, with strategies for real estate development. Governments can allocate land in a few large chunks for master-planned communities, or they can establish some general rules for land subdivision and development and encourage numerous smaller, independently initiated and controlled projects. Most often, in practice, urban form emerges from a complex combination of the two—and that is likely to be the future of the Hertzian landscape as well.

ELECTRONIC NOMADICITY

Gradually emerging from the messy but irresistible extension of wireless coverage is the possibility of a radically reimagined, reconstructed, electronic form of nomadicity—a form that is grounded not just in the terrain that nature gives us, but in sophisticated, well-integrated wireless infrastructure, combined with other networks, and deployed on a global scale. Leonard Kleinrock (one of the pioneers of the Internet) has defined this infrastructure as "the system support needed to provide a rich set of computing and communication capabilities and services to nomads as they move from place to place in a way that is transparent, integrated, convenient, and adaptive."[30] This requires, as Kleinrock notes, "independence of location, motion, computing platform, communication device, and communication bandwidth, along with general availability of access to remote files, systems, and services." The technical challenges of achieving this are significant, but they will gradually be worked out, and as they do, the social and cultural implications of electronic nomadicity will become increasingly evident.[31]

Other types of networks—transportation, energy supply, water supply and waste disposal—cannot operate wirelessly (or pipelessly), of course. But by providing efficient summoning and locating capabilities, wireless connectivity links our mobile bodies much more effectively to these more traditional resource systems. If you want transportation, for example, you can call a taxi or ambulance with your cellphone. If you want to know when the next bus is coming, you can look at a bus stop display that wirelessly tracks the vehicles in the system. If you want to find a nearby vacant parking space, a drinking fountain, or a public toilet that's open and salubrious, you will increasingly be able to do so with your portable wireless devices. If you discover a restaurant with a great special on the menu, you can call your friends. In these sorts of ways, wireless systems reduce search and uncertainty, and minimize the time required to get what we need.

Furthermore, wireless interconnectivity mobilizes things as well as people. In wired networks, things such as desktop telephones and computers need to be physically plugged in to operate; they have relatively fixed and stable locations. But in wirelessly interconnected systems, components just need to be within range. If you add miniaturization and self-configuration capability (just stick a component anywhere, and it works) to wireless interconnectivity, networked systems become fluid and amorphous.[32] They are less like rigid things on boards or in boxes, less like buildings or cities, and more like the camps of nomads—ready to move around and reconfigure, at a moment's notice, as required.

The cumulative effect of these transformations is profound, and will become more so as wireless technology continues to develop and proliferate.[33] Wireless connections of fixed infrastructure to wearable and portable electronic devices, and among miniaturized wireless devices, are now completing the long project of seamlessly integrating our mobile biological bodies with globally extended systems of nodes and linkages. As a result, functions that were once served by architecture, furniture, and fixed equipment are now shifting to implanted, wearable, and portable devices. And activities that once depended upon close proximity to sites of accumulation—of water, food, raw materials, bank vaults, library books, or files of business information—now rely increasingly upon mobile connectivity to geographically extended delivery networks.

In an electronically nomadicized world I have become a two-legged terminal, an ambulatory IP address, maybe even a wireless router in an ad hoc mobile network. I am inscribed not within a single Vitruvian circle, but within radiating electromagnetic wavefronts.

ACCESS RULES

This new form of nomadicity is, of course, very different from that of ancient hunter-gatherer bands, which were forced by the sparse distribution of food, water, and other resources to range over wide areas. For them, mobility was not only a necessity but also a developmental constraint. They were limited by what they could carry around with them. The very possibility of further economic, social, and cultural development depended upon sedentarization, the accumulation and protection of food surpluses, the consequent ability to support non-food-producing specialists at sites of accumulation, and increasingly specialized division of labor within densely populated cities.[34] Today, large-scale networks support division of labor and specialization on a global rather than a merely urban scale, the possibility of non-food-producing specialization depends upon network access, and that access is—increasingly—available ubiquitously and continuously. You can get whatever advantages you may seek from mobility—wider intellectual and cultural horizons, global business opportunities, access to unique talents and resources, collaboration among geographically distributed specialists, the stimulus of diversity—with far fewer of the traditional costs.

Under these post-sedentary conditions, access capabilities and privileges are more important than traditional forms of ownership and control of property.[35] If you are sufficiently wealthy and privileged, for example, you can now travel very lightly—with credit card and passport, some portable electronic equipment, and a carry-on bag; you can take full advantage of the world's highly developed network infrastructure to access whatever you want, wherever you may need it. If you are a knowledge worker, a personal library of books accumulated in your study may now be less useful than mobile network access and acquisition of intellectual property rights to online information.

The time-honored way to invoke access privileges is to proffer a physical token that you carry in your pocket, such as a talisman, passport, metal key, or plastic card, so that a gatekeeping system can look for a match. With computer systems you proffer some bits that you carry in your head—a password or PIN—by explicitly typing or swiping. Now you can simply equip yourself with a wireless transponder that instantly, automatically, and unobtrusively supplies ID bits when interrogated by an electronic gatekeeping device. Thus automobiles with EZ-Pass transponders can gain access to toll roads without waiting at tollbooths, and access to gasoline pumps through Exxon Mobil's Speedpass, which is based upon similar RFID (radio frequency identification) technology. With further miniaturization, RFID devices shrink to tags that are tiny enough to carry on key chains, to be sewn into clothing, or even to be implanted under the skin; wireless nomads can become continuously, automatically self-identifying.[36] You don't even have to stop to identify yourself; electronic devices can suck the RFID bits from your body or your vehicle as you pass by.

From the viewpoint of service consumers, ubiquitous access capabilities and privileges assure you of getting what you want when and where you want it. From the marketer's perspective, that very same connectivity provides opportunities (subject to any privacy constraints that may be applied) to continuously track your behavior, draw inferences from it, and anticipate your needs. (Step up to buy a hamburger, and automatically have it your way.) Whenever and wherever you electronically invoke your access privileges, your action is likely to be recorded in a globally centralized database and, eventually, used as a data point in computing your creditworthiness and likely future consumption patterns, and in pinpoint marketing strategies.

In other words, the post-sedentary world represents the ultimate abstraction and mobilization of exchange capability—and therefore, of wealth and power.[37] In simple barter economies, wealth does you little good unless you have your accumulated goods nearby and can physically hand them over in exchange for other things you may want. Coins and banknotes emerged to serve as more abstract and mobile exchange tokens and supported the development of more complex and geographically extended systems of trade. Electronic telecommunication networks took abstraction and mobilization still further—allowing wealth to become electronically manifest at Western Union offices, ATM machines, or credit card charge terminals at points of sale. Now, globally negotiable, electronic exchange tokens (which may be complex, metalevel abstractions such as mortgages, options, and derivatives as well as straightforward cash equivalents) can move wirelessly, and show up wherever a mobile device can get reception.

Conversely, post-sedentary space also redefines the condition of homelessness—the plight of being placeless and marginalized in a sedentary society. Today, it isn't fundamentally a case of having no fixed abode. It's one of having no access privileges. If you cannot afford or obtain such privileges, if you get blacklisted, if you simply lose your cards or equipment, if you forget your passwords, if you have a RFID tagectomy, or even if your batteries just run out, you are—like those clueless colonial explorers in the Australian desert—surrounded by inaccessible abundance.

NODULAR SUBJECTIVITY

For good and ill, then, I am a not only a networked, spatially extended cyborg, but also a post-sedentary one—not because I have been bionically rebuilt (apart from a couple of replacement teeth, I still rely upon my slightly worn original equipment), but because I am continuously connected, even when in motion.[38] I am inseparable from my ever-expanding, ever-changing networks, but they do not tie me down. Not only are these networks essential to my physical survival, they also constitute and structure my channels of perception and agency—my means of knowing and acting upon the world. They continuously and inescapably mediate my entire social, economic, and cultural existence. And they are as crucial to cognition as my neurons.

Sometimes these extended flow systems demand that I disclose the identity of my mobile body, but at other times they introduce identity impedance. When I stop to pee into a urinal, I engage a gendered network node and thereby make a declaration of gender. When I pass through an airport security checkpoint or operate an ATM machine, I must declare my name and document my right to use it. If I carry a RFID tag or submit myself (maybe unknowingly) to biometric scrutiny, the labels that my body carries are observable. But if I wear a mask and gloves, identifying labels are obscured. And on the Internet, as has endlessly been remarked, nobody knows I'm a ____. As my body extends artificially from its fleshy core, its gender, race, and even species markers may fade.[39] It may acquire multiple, sometimes contradictory aliases, masks, and veils. Its agents and avatars in particular contexts may be ambiguous or deceptive—as when I choose an electronic skin to represent me in a videogame. Its very location may become indeterminate, and it may hide itself behind encryption schemes and proxy servers.[40]

We need more than McLuhanist extensionism, and certainly more than unregenerate dot-com boosterism, to make sense of all this. It isn't simply that our sensors and effectors command more territory, that our webs of interconnectivity are larger and more dynamic, or that our cellphones and pagers are always with us; we are experiencing a fundamental shift in subjectivity.[41] As Mark C. Taylor has succinctly summarized, "In emerging network culture, *subjectivity is nodular* . . . I am plugged into other objects and subjects in such a way that I become myself in and through them, even as they become themselves in and through me."[42] I do not have a fixed identity, nor do I exist as a discrete individual. My spatial and temporal coordinates are diffuse and indefinite. My network extensions intersect and overlap with those of others.

Humanist sages could complacently claim that the proper study of mankind was man. That ex-cathedra confidence looks misplaced in a post-whatever, ex-net era; "mankind" and "man" are clapped-out categories, and the idea of "studying" (in a study?) seems increasingly anachronistic. For networked scholars like me—constructing texts on our wireless laptops, writing on the run, continually shifting and multiplying our geographic and electronic vantage points, carrying digital cameras, surfing the Web for our sources, tracing through networks of citations, cross-references, and hyperlinks, sending out agents and spiders, poking around in the metadata, and attending to streams of email and instant messages as we go—the pertinent preoccupation is the electronomadic cyborg.[43]

Many may mourn the passing of the (presumably pre-TCP/IP, pre-HTTP, pre-RFID) liberal humanist subject and its celebrants. Heideggerians and other critics of modernism may kvetch about totalizing technology and the allegedly alienating qualities of the wireless cyborg condition. Students of gender, race, and political economy may remind us (quite properly) that we are not all networked to the same extent, in the same ways. Defense and security specialists may worry (quite understandably) about the increasing destructive potential of network crackers and hijackers. Those who just want a simpler life may choose to unplug, and to live off the grid in Idaho. But for this particular early-twenty-first-century nodular subject, disconnection would be amputation. I am part of the networks, and the networks are part of me. I show up in the directories. I am visible to Google. I link, therefore I am.

4

DOWNSIZED DRY GOODS

Even the humblest of everyday artifacts can suddenly gain utility, claim new roles, and form new spatial patterns when they are radically downsized or lightened. Ryszard Kapuscinski, for example, has pointed out the effect of the "cheap, light, plastic container" on African communities. Once, the women had to carry water in heavy clay or stone vessels on their heads. These vessels were valuable, so the women stood in line with them, for hours, at the spring. Now, plastic containers are light enough to be carried by children and inexpensive enough to be left in line while you find some shade or go off to perform other chores. Kapuscinski comments: "What a relief this is for the exhausted African woman! . . . How much more time she now has for herself, for her household!"[1]

Ironically, the affluent now also get their water in lightweight plastic containers—with labels like Evian. In this case, lightening the container helps the distributor to bypass local water supply systems and to deliver a branded product from a great distance. From the consumer's viewpoint, lightness has a different value; it provides the product with portability, therefore adding to its appeal to travelers and recreationists. Lightness is what you make of it, in some particular context.

Since the beginning of the industrial revolution—and at an accelerating pace over the last few decades—designers have exploited new technologies to make things smaller and lighter. As they have crossed certain dematerialization thresholds, many different types of machines that were parts of the architecture have become parts of our bodies. And this has been crucial in production of the new nomadicity.

MINIATURIZED MACHINERY

Consider, for example, music storage and playback devices. Pianolas required piano movers, and pianola rolls took up a lot of shelf space. Gramophones were still bulky, but sufficiently portable to make it to the front in World War I.[2] Cassette tape players could go to the beach. The Walkman became wearable. And MP3 players have become smaller still, since they do not need to accommodate relatively bulky tapes or CDs. More and more music gets stuffed into smaller and smaller boxes. Once you might carry two or three tracks on your person; now you can carry thousands. The evolutionary path has led from heavy furniture to tabletop and desktop devices to handhelds and wearables.

As architects and product designers know, there is usually some critical subsystem that controls the size of the whole thing; get rid of it, or find some effective way to shrink it, and you can reduce overall size and weight. In structural systems, as Buckminster Fuller tirelessly pointed out, the main problem is with the compression members; find a way to replace as many of these as possible with tension members, and you get a much lighter structure. With the Diskman the difficulty was the diameter of the CD; no matter what you did, you could not make the player smaller than that—which meant you could not get it to fit in your pocket.

Furthermore, there are relationships between scale and material. It seemed natural to enclose early gramophones in wooden cabinets and to treat them as a new species of varnished furniture. As tabletop stereos emerged, the wood became increasingly residual; metal and plastic took over. And you would not dream of trying to fabricate a portable Diskman case in wood; you cannot work wood sufficiently effectively at that scale, and it does not have the right mass and strength properties. Sometimes the miniaturization of devices motivates a shift to new materials, sometimes the emergence of new materials and associated fabrication technologies enables a wave of miniaturization, and sometimes it's a combination of the two.

With cameras, the problem was film. The size of the negative controlled the dimensions of the optical system and the film transport mechanism. Shrinking film formats (and a move from glass to celluloid) accomplished a certain amount of miniaturization, but substitution of the CCD array, in digital cameras, decisively changed the rules of the game. Tiny, dense arrays allowed optical systems to shrink and completely eliminated the film transport mechanism. More subtly, substitution of electronic circuits for optical and mechanical means (particularly in the viewfinder system) changed the required connections and spatial relationships among camera parts and allowed them to be repackaged in denser and more compact ways. After a while, digital cameras were not only smaller than their predecessors, they did not look like cameras anymore—just as horseless carriages had stopped looking like carriages.

With laptop computers, you need a keyboard big enough to accommodate your fingers, fine motor skills, and a screen scaled to your visual field. Handhelds with palm-sized screens and ridiculously tiny keyboards are an uncomfortable compromise. But if you can substitute a retinal scanning display that paints a high-resolution image directly on the inside of your eyeball for the screen and a microphone hooked to a speech recognition system for the keyboard, you can shrink the whole thing to Rayban scale, and shift it from your knees to your nose.

Substitution of electronic connection (either wired or wireless) for mechanical linkages, optical mechanisms, or flows of materials also allows products to fragment and recombine. Their functions can be redistributed, in new ways, over portable devices, tabletop appliances, and fixed equipment. If you want to shrink and lighten something, you can therefore do so by offloading functions to some other location. In photography, for example, exposures have traditionally been made by handheld devices, with developing and printing operations consigned to centralized darkroom installations, and storage functions handled by albums and archives. The Polaroid instant process repackaged exposure, development, and printing into handheld devices— providing convenience at the cost of bulk. But digital photography provides almost unlimited freedom to recombine. The exposure device can be reduced to a lens and CCD array with a network connection, and it may be handheld, clipped on to another device such as a PDA or cellphone, or fixed to a wall. Images may be stored in a portable device, a desktop device, or a network server. And printers may be located wherever there is a network connection.

Sometimes, such recombinations open up new opportunities. Instant photography provided the possibility of discussing and evaluating images within the contexts in which they were produced rather than at later times and in different places. Similarly, the integration of digital cameras with cellphones provided callers with the opportunity to show what they were talking about instead of describing it verbally.

MICROFABRICATION AND MEMS

High-resolution fabrication technology provides yet another way to downsize useful devices. This has been most dramatically demonstrated in the design and production of electronic circuits. The vacuum tubes used in early computer circuitry were, unavoidably, bulky and hot. The transistors that soon replaced them were smaller and cooler and could be packed much more densely. Then semiconductor technology put the explosive exponent into Moore's Law of silicon scaling. In the 1950s, portable radios with half a dozen transistors seemed miraculous; by the turn of the century, postage-stamp-sized computer chips with 100 million transistors were commonplace.

Microfabrication typically begins with a macroscopic element, such as a wafer of silicon, and creates complex structures, such as integrated circuits, by precisely removing or depositing material. As the technology has advanced, the minimum dimensions of elements in these structures have shrunk from tens of micrometers to tens of nanometers. This progression will reach its limit when elements get down to a couple of nanometers—the size of an atom—but this is not the end of the technological line for microfabrication.[3] As the race to this limit nears its end, emphasis is shifting to invention of new types of microscale structures and systems.

Already, microfabrication techniques have been extended and generalized from electronic circuits to microfluidic systems with tiny channels, reservoirs, valves, and nozzles to replace the glass tubes and beakers of traditional chemistry laboratories, and thus allow analysis of much smaller samples. They have also been employed to produce waveguides for optical and radio signals. The name for such structures, microelectromechanical systems, is bigger than they are—so it has mercifully been shortened to MEMS.[4]

More surprisingly, MEMS can have moving parts such as switches and valves, vibrating cantilevers, and tiny gears and mechanical linkages. This enables MEMS to function as sensors that transduce some aspect of the environment into electronic data. They can, for example, serve as pressure sensors, microphones, accelerometers, and angular rate sensors. They can be employed to detect visible and infrared light. And they can become "laboratories on a chip" to detect chemical and biological agents.

Conversely, MEMS can function as actuators that transduce information into useful physical, chemical, and biological effects. They can, for example, emit light or radio frequency (RF) signals, adjust microscopic mirrors to direct signals in fiberoptic systems, and serve as motors to propel microscopic vehicles and robots.

In the early days of microfabrication, microchips usually served as the intelligence in macroscale devices. The personal computer of the 1980s defined the genre; it was a microchip, surrounded by a lot of other stuff that provided the power, and did the sensing and actuating, in a large box. Through the 1980s and 1990s, microchips were embedded in a widening array of macroscale systems, from household appliances to automobiles and aircraft. Now, as MEMS technology develops, the other stuff can often shrink as well. This is opening up new design possibilities. MEMS devices can function as insect-sized autonomous systems within the human body, and in other contexts that demand extreme miniaturization. Batch-fabricated, inexpensive MEMS can be scattered around like grains of wheat, painted onto surfaces, or mixed into materials like concrete. They can be arrayed to form intelligent skins that sense changes in their environments and respond appropriately. And they can intercommunicate and coordinate wirelessly to form systems with distributed intelligence.

THE RISE OF NANOTECHNOLOGY

Beyond micro there is nano—the world of atom-by-atom or molecule-by-molecule construction of devices and systems with key dimensions measured in billionths of a meter. The idea goes back to a famously inspiring talk by Richard Feynman, in 1959, entitled "There's Plenty of Room at the Bottom."[5] In the late 1980s, K. Eric Drexler set off a new wave of interest with his speculative book *Engines of Creation*.[6] Little more than a decade later, there was a heavily funded National Nanotechnology Initiative in the United States, and similar efforts in other parts of the world.[7] Science and technology magazines were publishing regular—sometimes breathless, sometimes critical—overviews.[8] And Michael Crichton, in his technothriller *Prey*, had seized upon the idea of nasty, self-reproducing nanoparticle swarms as a new way to scare his readers.[9]

Not only are nanoscale widgets smaller than their microscale cousins, they also behave differently. Quantum physics kicks in. Atomic forces and chemical bonds dominate. Surface-to-volume ratios are large—often yielding useful chemical and biological properties. Issues of strength and proportion, power-to-weight ratios, friction, heat dissipation, and durability and reliability tend to work out differently than they do at larger scales. You have to worry about tiny moving parts banging into relatively large air molecules. Down there at the bottom, designers must play by new rules.

In 1981, the introduction of the scanning tunneling microscope opened up the possibility of imaging and manipulating single atoms on surfaces. Since then, nanotechnologists have employed a variety of scanning microscope techniques—in particular, atomic force microscopy—to push atoms around like Lego blocks. This provides a way to handcraft interesting nanostructures, but fabrication of such structures in useful quantities has turned out to require an eclectic mix of techniques drawn from physics, chemistry, materials science, mechanical engineering, electrical engineering, and biology. At nanoscale, many of the traditional boundaries among these fields disappear.

Where microfabrication depends upon top-down imposition of patterns on material, nanoscale fabrication processes may work by bottom-up self-assembly. As in biological systems, structures are automatically built up from atomic- and molecular-scale units, then substructures are assembled into larger and more complex units, and so on. If you want to build very complex structures in this fashion, you have to find ways to minimize errors, and to correct errors automatically when they occur.

At nanoscale, the possibilities of molecular electronics and quantum computation begin to open up. Nanoelectronic circuits might be built from molecular "wires,"[10] or from quantum dots—wireless structures built up from electromagnetic "boxes" holding discrete numbers of electrons.[11] Computer memories and displays might be constructed from carbon nanotubes.[12] Complete "computational particles"—working together as amorphous computing systems—might become small enough to sprinkle like dust, float like pollen, or be injected into the bloodstream to serve as diagnostic devices.[13] And chemical and biological sensors might assay single molecules.

NEMS (nanoelectromechanical systems) might incorporate molecular moving parts. Already we see microscopic motors, gears, chains, pumps and accelerometers, bug-sized robots, and coin-sized turbines, and the Web has numerous picture galleries of designs for nanometer-scale mechanisms. Nanoscale machines can even join the wireless world. It is now possible to attach a nanocrystal antenna to an individual DNA biomolecule, so that it can be controlled remotely by radio signals.[14] It can twitch reversibly on command—to function, for example, as a tiny actuator or switch, or to change its expression within a biological system.

REFRAMING DESIGN TASKS

Extreme miniaturization is usually portrayed as a path to higher speeds, greater efficiencies, more economical use of materials, and lower costs. But it also provides a way of squeezing more functions into smaller packages, so moving them closer to the body (or even inside the skin) and freeing them from fixed locations. Sophisticated computer graphics functions, for example, first became available on terminals attached to mainframe computers, then on desktop workstations, then on portable game consoles and laptops, and now on MEMS-based retinal scanning devices. Artificial hearts began as bulky, bedside machines in hospitals, then they shrank sufficiently to become implants.[15]

In the nano era, possibilities expand further. Richard Feynman imagined putting a tiny, robotic heart surgeon inside a blood vessel, thus dispensing with surgical suites.[16] And the irrepressible Ray Kurzweil has even proposed sending billions of nanobots into the brain to replace virtual reality goggles; "If you want to be in real reality, the nanobots sit there and do nothing, but if you want to go into virtual reality, the nanobots shut down the signals coming from my real senses, replace them with the signals I would be receiving if I were in the virtual environment, and then my brain feels as if it's in the virtual environment."[17] It's like Alzheimer's, but with active, benevolent nanobots rather than passive, destructive platelets. From a network control viewpoint, it makes sense; instead of replacing signals to a couple of nodes (eyeballs) at the very edge of the neural network, go for lots of nodes at the core of the network.

This shift back to the body has also altered the context and framing of design tasks. Wall and desktop telephones, for example, have long been assimilated to the tradition of mechanical and electrical appliance design—that of clocks, toasters, coffee grinders, and stereos. Their designers are schooled in the minimalist, universalist ways of the Bauhaus and Ulm, and the most elegant exemplars find their way to the industrial design section of MoMA. But cellphones are increasingly conceived of as personal accessories—much like wallets, handbags, shoes, hats, neckties, and spectacles. It is turning out that gender, age, and status markers are important; a senior, male financial executive usually wants something that goes with his suit, while a Japanese teenage girl may prefer Hello Kitty. When phones migrate from walls and desktops to pockets, they also move into the domain of fashion design and marketing—and their forms and styles, like those of clothing, proliferate endlessly. When you begin to wear them as emblems, rather than carry them as tools, they play a different cultural role.

With computers, the shifts have been even more dramatic. Mainframes were designed as large-scale items of industrial equipment, and at their best—in the hands of Charles Eames, for example—achieved a tough, hard-edged, machine-age clarity of form.[18] They were often put on display in special, glass-enclosed rooms. The bulky computer workstations of the 1970s and 1980s were medium-scale wheeled furniture—not too different from writing desks, pianos, and treadle sewing machines, but styled for laboratory rather than domestic environments. PCs evolved from clumsy beige boxes to sleekly specialized, variously colored and shaped versions for offices, classrooms, and homes.[19] Now that they are fading into history, after a life of approximately twenty years, they look increasingly like surrealist constructions—the chance encounter of a typewriter and a television on a desktop. Portables started out mimicking luggage (right down to the handles and snaps), then appropriated the imagery of books that could open, close, and slip into a briefcase. Even smaller versions snuggled into pockets and handbags, like cigar cases, hip flasks, and makeup compacts. Next, as components became tinier, and as designers realized that parts could be interconnected flexibly rather than packed into rigid plastic or metal boxes, computers became conformable to the contours of the body. They had evolved from heavy machinery into close-fitting wearables; you could begin to imagine wiggling into them like gloves, folding them into pockets like handkerchiefs, or sporting them like neckties. Ultimately, you might think of them as smart, barely visible particles.

Once, designers separated their domains by scales and associated functional categories; circuit designers and nanotechnologists operated in the nanometer to millimeter range, product designers went from millimeters to meters, architects typically dealt with details at millimeter scale and overall building dimensions of tens or hundreds of meters, while urban designers and civil engineers might work on infrastructure and land use systems extending over many kilometers. Today, such scale chauvinism makes little sense. The solution to a given design problem might be found at any scale or combination of scales—and an increasing amount of functionality that once resided in large, immobile structures and machines is now squeezed into portable, wearable, and even molecular devices.

It makes even less sense to draw sharp distinctions between nonliving and living systems. As biology, materials science, mechanical engineering, and electronics all get down to the molecular scale, they deal with the same types and sizes of structures, and there is a growing crossover of interests and goals. As biologists engage ideas of modular recombination, splicing, and cloning, they begin to think like designers. Conversely, as designers tentatively embrace concepts of emergence, self-organization, self-assembly, and self-replication, they start to sound like biologists. Increasingly, the CAD console meets the wet lab, and the circuit shop keeps company with the chemistry bench.

MULTIFUNCTIONALITY

Growing reliance upon small-scale systems—particularly miniaturized, portable electronics—has also produced rampant hybridization of devices. Not so long ago, for example, telephones were desktop or pocket devices for audio communication, cameras were optical/mechanical/chemical devices for picture taking, and GPS navigation systems were bulky items of equipment for boats and airplanes. By about 2002, though, all of these devices could be squeezed into the same portable, electronic box—and their combination opened up a surprisingly useful new possibility; you could take a picture and instantly transmit it, along with a map of the place where the picture was taken.[20] Servers could begin to accumulate image databases—with automatic indexing by time and date, location, and author—from multiple, mobile, remote sources.

Similarly, cellphones and PDAs, which had arrived upon the scene as separate boxes, began to fuse into one.[21] This convergence was prompted, in part, by the competition for pocket and handbag space; why carry around two boxes when you can get what you need from one? And it was also motivated by the search for economies; why double up on power supplies, processors, display screens, and keyboards? But its most important consequence was the convenient integration of functions that had hitherto been separate; why keep your address book and your dial-up device in separate, disconnected boxes?

There is, however, a crucial space-time tradeoff to consider. When a device such as a Swiss Army knife or a PDA provides access to many different functions, you can only make use of one of these functions at a time, and you have to switch from mode to mode—from the knife blade to the corkscrew, for example, or from the address book to the calendar. Conversely, a spatial array of single-purpose knives, corkscrews, and so on takes up more space, but there is no time-wasting and potentially confusing mode switching, and you can store each special device in its context of use—the corkscrew near the wine rack, and so on. Where space is scarce and there is little or no fixed infrastructure to rely upon, as in a hiker's backpack, multifunctionality and form factor minimization tend to win. Where space is fairly tight, as in a tiny city apartment, multimodal devices such as sofa beds may still make sense. But where there is plenty of room and a lot of fixed infrastructure, as in a large suburban house, it is far more convenient to provide special-purpose devices in their contexts of use—beds in bedrooms, sofas in living rooms, corkscrews at bars, and knives in kitchens.

It makes a difference if you can switch modes easily; unfolding a sofa bed for sleeping is aggravating and time-consuming, but picking a function from a menu on a PDA is not so bad. Context-sensitivity, if it can be reliably achieved, is even better; a really smart portable device might know where you are and what you need to do there, and adjust its current mode accordingly.

Miniaturized, mobile devices sometimes allow us to save time by performing tasks while we are in motion. Most of us, for example, have no difficulty listening to the radio while jogging or driving. But our capacity to pay attention is limited, so driving while talking on a cellphone is riskier. When you need to use keyboards and view video screens, it is wise to stop, sit down, and give them your undivided attention. As broadband wireless connections deliver fatter streams of bits to the mobile body, attention management will become an increasingly crucial design issue. The mechanisms may be very simple, as when voicemail or TiVo allow you to defer attention to data streams until you are ready. Or they might depend upon sophisticated context-sensitivity—enabling a cellphone or automobile navigation system to interrupt you only when it is safe to do so, but to be bolder when the message is really urgent.

As devices become smaller, as software takes over more functions from hardware, and as space-time tradeoffs are critically reevaluated, traditional functional categories may no longer hold. Electronic devices may readily be assembled into unprecedented combinations that provide hitherto unavailable functional mixes. These assemblies may be stuffed into compact packages, or they may be constructed

through dispersed network connections. Through wireless interconnections, these functions may be divided, in whatever ways turn out
to be most convenient, between smaller, wearable devices and larger
elements of immobile infrastructure. And, by virtue of their embedded
electronics, objects that have long performed traditional functions—
from items of clothing to sheets of wallboard—will acquire increasingly important ancillary functions.

Where efficiency matters, as in the layout of a chip, form follows
function in rigorous fashion; sophisticated optimization techniques are
used to minimize the distances that electrons must move through surfaces, to pack components as densely as possible onto scarce chip real
estate, and to assure that heat is effectively dispersed. Conversely, at
scales and speeds where the lengths of the wired or wireless linkages
among standard electronic components have trivial effect upon performance, designers have enormous freedom to shape electronic assemblies into arbitrary sculptural forms, to mold them to the contours of
the body, to conceal them within other objects, and so on. During the
1980s, the Department of Industrial Design at London's Royal College
of Art championed this freedom, and many innovative designs for
electronic products emerged—most notably, perhaps, Daniel Weil's
designs for transistor radios. More recently, designers of electronic toys
have exploited it in increasingly imaginative ways—dolls converse
electronically, toy vehicles acquire sophisticated electronic functionality, electronic dogs and other pets learn from their environments and
respond to care; Lego puts electronics into modular building blocks,
and Tod Machover's *Toy Symphony* blurs the line between electronic
play and performance.

BACK TO THE BODY

All this has intensified interest in the scarce real estate of skin surface
and its immediate surroundings. When timepieces resided in clock
towers, they competed for central urban sites, but when they shrank
to watches, they began to compete for wrist space—scarcer (at least
by convention) on female wrists than on male ones. When Marconi set
out to build his Atlantic radio telegraph station, he first had to find
an industrial-scale chunk of real estate on Cape Cod, but when we

decide to carry cellphones, we have to find space in our limited pocket, belt, or handbag real estate. Topography established the context for Marconi's design; anatomy does so for a cellphone design.

Where architects have traditionally responded to human needs by allocating square footage for mechanical and electrical systems, furniture, and equipment within the rigid, large-scale fabric of buildings, cyborg couturiers are now doing so by locating miniaturized devices within the smaller-scale, more flexible fabric that clothes us.[22] The microterrain immediately surrounding our bodies is providing habitats (mostly with very limited carrying capacities) for new, electronic species, which may be classified according to their sizes and shapes (known in the trade as their form factors), their modes of attachment to the body, their degrees of conformability to the body, and their degrees of visibility. You can think of these species as electronic parasites that both depend upon their hosts and provide benefits to them. And these parasites are evolving rapidly as they compete for the available niches within this up-close-and-personal terrain.

Traditionally, these niches and species have been explored most rigorously by designers of gear for foot soldiers—who need to be as effectively equipped as possible, but who cannot be expected to lug too much weight. A Roman centurion carried around about forty-five pounds, but a modern soldier may be burdened with three times that. Not surprisingly, then, the U.S. Army has been quick to establish an Institute for Soldier Nanotechnology, focusing on "threat detection, threat neutralization (such as bullet-proof clothing), concealment, enhanced human performance, real-time automated medical treatment, and reduced logistical footprint."[23] According to the director: "This will be achieved by creating, then scaling up to commercial level, revolutionary materials and devices composed of particles or components so tiny that hundreds could fit on the period at the end of this sentence." Reversing their usual tendency, weapons systems planners are beginning to think small.

ELECTRONIC PARASITE NICHES

Some emerging, miniature electronic species find their niches—at least initially—in the strap-on, backpack systems such as the packs of hikers and foot soldiers, and the wearable equipment of scuba divers. These systems typically consist of fairly large, rigid elements linked by flexible fabric, leather, or hinges so that they can conform reasonably closely to the contours of the back. The more loosely strapped, shoulder-hung variants can accommodate smaller objects, such as cameras. This part of the terrain has a lot of carrying capacity, but it produces cumbersome, clumsy appendages (which are particularly irksome indoors and in confined spaces), so it is best avoided whenever greater miniaturization provides the opportunity.

The possibility of gripping large objects of arbitrary shape in the hand provides another attractive parasite niche. Historically, it has been occupied by the luggage of travelers, by large weapons—from spears to shotguns—or by specialized mechanical devices like portable typewriters. Generally you can only carry one or two things, so competition for this niche is intense. In the latter half of the nineteenth century, the winner was often the gentleman's walking cane, which therefore acquired an astonishing range of specialized secondary functions—weapon cane, tippling cane, pooper-scooper cane, forked cane for trapping snakes, bicycle pump cane, seat cane, gun rest cane, cigarette case cane, gas lighter cane, watch cane, spyglass cane, zither cane, flashlight cane, cologne cane, and many more.[24] Today the victor in this niche is commonly a laptop computer with a handle or carrying case—a device of high functionality that cannot easily move into smaller-scale habitats because of the need for large screens and convenient keyboards. Like the cane, it can accommodate numerous secondary functions—now provided by plug-in peripheral devices, such as DVD drives for playing movies. The principle is the same, though the form factor (driven by the primary function) is different. But the laptop's victory is a tenuous one; handheld objects are always in danger of being put down for something else.

The traditions of clothing design have established an important niche for electronic parasites that can be slipped into pockets or hung in pouches and holsters on belts. Since bulging pockets are uncomfortable, and tend not to be regarded as a fashion plus, this niche imposes severe restrictions on the length, width, and thickness of objects. And, since pockets are flexible containers that may be subject to stresses as the body moves, it is an advantage for objects competing for pocket space to be flexible, conformable, and resilient. So far, the most successful electronic invaders of pocket space have been cellphones, PDAs, and electronic cards of various kinds. Devices that can squeeze into this niche can become almost as inseparable from us as our underwear. As miniaturization reduces more electronic devices to pocket size, and as new polymer-based technologies enable flexible batteries and circuits, the competition for pocket space is likely to intensify.

Still smaller electronic parasites may be sewn to clothing like buttons, pinned on like badges, strapped on like watches, or directly attached (with or without body piercing) like finger rings, navel jewelry, and ear studs. Cameos and lockets may run video loops instead of displaying static images, and sparkle may be provided electronically rather than by the cut of a gem. At this scale, conformability and flexibility matter less; jewel-like devices can, without producing discomfort or inconvenience, be rigid and quite freely shaped. Furthermore, systems of rigid, jewel-sized elements can be connected flexibly (like strings of beads) to make much larger, conformable constructions.

Let's not forget teeth. If you have to get a gold tooth or ceramic crown, why not pack it with electronics? If your teeth carried an RFID tag, you might make purchases or open hotel room doors by flashing a smile. Maybe a memory filling would be a good, safe place for your crucial medical records. And, if you squeezed a wireless speaker into a molar, you could take advantage of the fact that your jawbone efficiently transfers sound and eliminate the earpiece of your hands-free cellphone. The generalization to nails and lashes is obvious.

And finally, of course, electronic parasites are increasingly capable of getting under your skin. Although they have been framed culturally in very different ways, the practices of body piercing and subdermal implantation are not far apart technologically. You can, for example, have a rice-grain-sized RFID chip injected by syringe to provide purchasing power and location-tracking capability.[25] Deeper within the body, we will increasingly find cochlear implants, internal defibrillators, and miniature devices incorporating sensors and transponders capable of measuring blood pressure and other conditions transmitting data wirelessly to external receivers. These devices are more permanently attached than external wearables, and you do not have to remove them to take a shower.

Some of these proliferating parasites can attach themselves wherever there is room, but others require particular anatomical contexts. Thus there are growing, highly specialized genres of miniaturized eyepieces, earpieces, mouthpieces, and even nosepieces—all of which may, on occasion, be integrated into specialized masks or helmets. Wrists are good places not only for displaying time but also for other small-screen information. Shoes can provide convenient, well-engineered housing not only for batteries but also for generators that harvest the energy of footsteps to recharge them—and maybe even for energy storage and actuators that would enable you to leap tall buildings in a single Superman bound. And complete exoskeletons, composed of nanoparticles and electroreological fluids, promise both protection and the superhero accoutrement of a stiff, high-powered "forearm karate glove."[26] Of course, you don't have to be Stan Lee to imagine Marvel Comics scenarios other than those sketched by researchers in search of military funding; imagine street protestors who could effortlessly vault over riot police, equestrian and motorcycle gear that stiffened into a protective carapace when you were thrown off, and exoskeleton extreme sports.

SMART THREADS

Since the accommodation available at any point in intradermal and near-extradermal terrain is limited, and since there are efficiencies to be gained by centralizing rather than duplicating common functions, there is a growing need for network linkages among the electronic devices distributed around and within the body. It may make practical sense, for example, to centralize power supply instead of equipping each device with its own batteries or generators. Or there may be multiple, parasitic power generators—sucking in kinetic, thermal, light, and radio frequency energy at various bodily locations—to create a miniature power supply grid. This, then, introduces the need to drape the body with power cords, or—perhaps more elegantly—to weave power distribution circuitry (maybe composed of conductive polymers) into clothing.

For data networking there are more options. The links may be wired, as with the connections between pocket telephones or music players and earphones—the wires loosely draped or running elegantly through seams or zippers. Or links may be wireless, allowing anatomical logic rather than contiguity requirements to dictate disposition of functional elements around the body—relatively bulky power supplies and processors in pockets, audio output in the ears, video displays in handhelds, on wrists, or integrated with spectacles, sensors wherever they are needed, and so on. They may even be run (harmlessly) through the body itself.

Being smartly turned out will take on a whole new meaning as clothing fibers and fabrics acquire more active functionality and become increasingly programmable.[27] They might, for example, expand to keep you warmer in cold weather, open up to provide more ventilation in hot weather, tighten and decrease porosity to become waterproof, change color on command, and stiffen to provide protection in the event of accident or attack. By incorporating microcapsules of phase-change material, fabrics might absorb energy to cool you down when you're sweaty and expel heat to warm you when you're chilly. Gloves, socks, and tights might be programmed to teach dancers and athletes by applying tactile prompts, or to diagnose injuries by detecting changes in gait. And "accessorizing" will mean adding new devices to your network.

Thread with extended functionality will open up new possibilities for the ancient crafts of weaving and embroidery. Electrically conducting thread will allow circuit embroidery. Weaves of actuating thread will allow dynamic, programmable shaping and fitting. Threads that can vary their color will open up the possibility of animated tweeds and plaids. A programmable tie, woven from smart thread, might knot itself automatically and download patterns from the Internet.

Just as portable wireless devices are connected to nearby transmission and reception points, networked bodies may become mobile subnetworks of larger networks—maybe using cellphones, wireless PDAs, or button-scale transmitters and receivers to establish the necessary external links. They may incorporate RFID tags to electronically provide information about themselves. And, since the body itself produces low-powered electromagnetic radiation, it may function as a naked network node—enabling, for example, remote wireless monitoring of heartbeats.[28] The ancient, mystical idea of the body's ineffable aura takes on, in this context, precise engineering meaning.

Some of this will probably turn out to be miscalculated science fiction, and some of it will soon seem banal, but it's the repetition of a familiar theme that matters. Before the industrial revolution, buildings were mostly big, dumb boxes; then they acquired increasingly sophisticated mechanical, electrical, telecommunication, and control systems. Now, by taking advantage of electronics and nanotechnology, the rag trade is following in the footsteps of the construction trade.

AMBULATORY ARCHITECTURE

As the functionality of small things increases, they are insinuating themselves, like resourceful ticks and fleas, into increasingly intimate spaces. The traditional roles of clothing systems—thermal protection, waterproofing, impact protection, identity signaling, and so on—are being rethought and addressed in radical new ways.[29] And the elements of clothing systems are acquiring growing repertoires of new functions.

So designers are asking themselves some new kinds of questions. What might you now put in your pockets, wear on your belt, or carry in a backpack? What sorts of functions can you fit into jewelry, and how should jewelry express these? How might underwear and implants work together? What can be fitted, and what must hang more loosely? How much useful electronics can you jam into your shoes or your hat? What should be implanted semipermanently, what should go into inner garments, and what is best accommodated in easily shed outer layers? How can electronic earpieces, eyepieces, and mouthpieces be assimilated to traditions of facial adornment? Can you make use of rings or gloves to sense finger gestures—maybe replacing keyboards? How should wearable devices respond to bodily movement, changes in surrounding conditions, and emergencies? Might programmable, animated tattoos and makeup serve your display needs? What has to be waterproof? Where should we run the bodynets, and what should we run through them?

The ongoing shift of functions from urban and architectural to bodily real estate inverts some familiar customs and rituals. You once slipped into a telephone booth to make a call (and Clark Kent did so to change his costume), but you now slip a cellphone out of your pocket. Playing music through a stereo system at a party is a social gesture, but playing it through a Walkman is a way to withdraw. In a movie theater you look for a good seat and orient your eyes to the screen; with a portable display, you sit down anywhere and arrange the screen in front of your eyes. We are evolving our manners and social conventions in response—learning to avert our eyes from our seatmate's laptop screen and responding to "no cellphone" signs as our forebears responded to "no smoking" and "no spitting."

Where walls once established relatively clear and stable boundaries among social settings, mobile devices create unexpected, and sometimes difficult-to-manage juxtapositions. When your cellphone rings, there is a potential conflict between the behavioral requirements of your current physical setting and those of your electronic environment. You might choose one over the other, thereby alienating your companion or your caller, or the parties to the call might pass the phone around to their companions, using a simple ritual to bind distant social settings temporarily together. You might also scramble social settings, to the possible discomfort of others, by taking work calls in domestic or recreational settings, or personal calls in formal work settings. And you might unconsciously let the demands of one setting dominate those of the other, making motorists wish that you would shut up and drive, lecturers wish that you would look up from your screen and listen, and callers wish that you weren't simultaneously emailing someone else.

And all this, of course, transforms the ancient logic of threat and defense; the suicide bomber, with a small but powerful explosive device strapped inconspicuously under his shirt, is dramatic testimony to that. Airport security managers have become very interested in shoes. Miniature, self-actuating weapons—from high-powered letter bombs to anthrax spores—may be sent through the mail and might even be attached to insects or nanorobots. And it no longer suffices to frisk for guns and knives; checkpoints must rely increasingly upon sophisticated electronic detection of concealed devices and tiny traces of chemicals and biological agents. The boundary is blurring between systems that render the body transparent for medical purposes, such as x-ray machines, and those that do so for security.

As miniaturization continues, and as more and more functionality migrates to the body, literally off-the-wall (and into the skin zone) design moves will cease to seem so strange. We will indeed approach the condition of "walking architecture." We will be forced to abandon the macho prejudice that soft "fashion" is frivolous, while hard "construction" is serious. The functions of flexible, mobile clothing will be tightly integrated with those of rigid, fixed infrastructure. The IEEE will meet *Vogue*, and MIT will find common ground with FIT; the subtle skills of the clothing designer will be drawn together with those of the electronics engineer and the nanotechnologist to redefine radically the role of the first few millimeters surrounding our perspiring biological perimeters.

5

SHEDDING ATOMS

There are ultimate limits to the miniaturization of machinery; you eventually get down to atom-by-atom assembly (as imagined by the nanotechnologists), and there you have to stop—unless you consider atoms themselves as machines for producing subatomic particles. But with pure information products it's different; you can shuck bits entirely from their material substrates. You can store, reproduce, and transmit them as completely dematerialized patterns of electromagnetic energy.

One effect of this (in combination with the miniaturization of electronics) is to compress storage. We can now keep almost unimaginable quantities of digital information on servers the size of domestic appliances. And we can effortlessly carry entire libraries around with us in compact, lightweight, portable devices; pocket MP3 players, for example, can now contain the sorts of music libraries that once took the form of shelves of vinyl disks.

A second effect is to transform the character of information products. Digital texts, images, and other artifacts begin to behave differently from their heavier, materially embedded predecessors. They become nonrival assets—they are neither depleted nor divided when shared, they can be reproduced indefinitely without cost or loss of quality, and they can be given away without loss to the giver.[1] Thus they can support the dissemination, application, and creative recombination of innovations on a massive scale—provided, of course, that intellectual property arrangements do not get in the way.

A third effect of information dematerialization is to revolution-ize logistics. Instead of relying upon physical transportation systems, with their strict speed and capacity limits, we can pump bits through wires at the speed of light. As network bandwidth increases, we can do this in larger and larger quantities. And as the Internet has so dramatically demonstrated, we can employ computer intelligence to manage complex flow and delivery processes automatically.

And yet another effect—a particularly powerful one in combi-nation with all of these—is to enhance the mobility of information producers and consumers. Increasingly, we can download whatever we want, whenever we want it, to portable wireless devices. Conversely, we can upload products that we create while on the move. This chal-lenges the very idea of a fixed workplace.

Dematerialization delivers us from servitude to places and things—and, indirectly, from domination by those who control places and things. It undermines the regime of physicality. It constructs a new form of power and simultaneously provides a new way to resist power.

DEMATERIALIZED TEXT

Take text, for example. When it was inscribed in stone and clay, it didn't move very much; to gain access, you traveled *to* it. Then, when it shifted to lightweight sheet materials—papyrus, parchment, and paper—it began to circulate. Medieval monasteries became nodes in manuscript production, distribution, and consumption networks. With cheaper and more plentiful paper, printing, more efficient and reliable transportation, and mass literacy came large-scale, high-volume mail networks. Next, the telegraph network eliminated the paper substrate (over the long-distance legs of communication systems, at least), and demonstrated that short, electronically encoded strings of characters could move far faster than the swiftest messen-ger. Finally, digital storage and processing, ASCII coding, packet switching, and high-bandwidth electronic channels enabled the high-speed transmission of very large quantities of text. Today, through email, instant messaging, and the Web, text mostly comes to *you* in completely dematerialized form.

You can see the effects of all this in the varying forms that libraries have taken, and the social and intellectual roles that they have played. Before texts were mobilized, large concentrations of manuscripts functioned as powerful magnets for scholars and centers of scholarly communities.[2] The great Library of Alexandria, for example, attracted scholars from all over the Hellenistic world. The more extensive and renowned its collection of hundreds of thousands of scrolls became, the more scholars needed to be there. The surrounding community became increasingly rich in intellectual talent, and the daily interactions among members of that community multiplied the benefit—often yielding new manuscripts that ended up back in the library. This was, of course, the beginning of the modern idea of the university. Millennia later, the mechanism was still working just fine; when the more scholarly protagonists of *Ulysses* got together after lunch in the National Library at the Heart of the Hibernian Metropolis, the books brought them there and the intellectual interchange followed.

Well into the age of print, it remained slow, expensive, and difficult to build large book collections, so locations fortunate enough to have them sustained a comparative advantage. The buildings that housed them and celebrated their importance were often conceived as central landmarks—the library at the very heart of Venice, the great rotunda and dome at the focus of Thomas Jefferson's University of Virginia, the proud book towers at Cambridge and Yale. Within these landmarks, card catalogues near the entrances provided compact, central representations of the collections that could be surfed for guidance before venturing into the vast book stacks beyond.

But as books became cheaper and more numerous, and distribution systems improved, this traditional comparative advantage began to erode. Even small, isolated communities could now have respectable local libraries; these could not compare to the great central collections, but they sufficed to launch many a scholarly career. Local schools, colleges, and universities could develop their own collections. And individuals could stock their private bookshelves. The system for providing access to accumulated texts devolved into an increasingly multinodal, decentralized network of local storage and redistribution points.

As accumulations grew within this system, so did pressure on storage and transportation systems.[3] One response was to improve the engineering of bookshelves and bookstacks. Another was to make books smaller and lighter; the pocket book and the paperback emerged. Yet another was to compress more radically by transferring text from paper to microfilm.[4] By the mid-twentieth century, Vannevar Bush could see where this drive toward miniaturization and dematerialization was leading. In his remarkable vision of the "Memex," he imagined individual microfilm libraries, stored within desk-sized storage and access devices, which would provide individual scholars with inexpensive, immediate, local access to the resources of major collections.[5] And implicitly, he raised questions that were to reverberate, with ever increasing urgency, in the coming decades. To what extent should we treat bound books as valuable objects in themselves, justifying the allocation of space, time, and money for their preservation? And to what extent should we just strip out their information content, transfer it to more compact and mobile media, and toss away the carcasses?

It turned out that microfilm was not really up to the task, but the CD-ROM and the personal computer were. In 1972, on the Irvine campus of the University of California, the Library of Alexandria was reborn in digital form. The *Thesaurus Linguae Graecae* (*TLG*) project began, with the ambitious goal of producing a database containing *all* ancient Greek literature.[6] By 2001, the database contained virtually everything surviving from the period between Homer (eighth century B.C.E.) and 600 C.E., plus numerous additional texts from the Byzantine and post-Byzantine periods. David Packard's Ibycus system, and other software, provided sophisticated search and retrieval capabilities. It added up to more than 80 million words, and it was all available to scholars on CD—an astonishing feat of comprehensive accumulation, compression, and distribution that has transformed the daily practices of classical scholarship throughout the world.

The *TLG*, in its initial form, was highly miniaturized but still not completely dematerialized; the plastic disk of the CD remained and required physical transportation to sites of use. But packet switching, the Internet, and eventually the World Wide Web, allowed the remaining atoms to be scraped off.[7] Today, if you have a PC, a network connection, and a license, you can get the *TLG* online. You can download any ancient Greek text you want. Finally, Alexandria comes to you.

Encyclopedias have, with astonishing rapidity, followed a parallel developmental trajectory. Until the PC came along, the *Encyclopedia Britannica* was a bulky, heavy, expensive set of printed volumes. Then the Encarta encyclopedia shrunk itself onto a CD. Now the *Britannica* has dumped those high-priced atoms (for which there is no longer a market) and has gone online—supporting itself through advertising and ancillary service revenues. And in specialized professional fields, subscription-supported online databases, such as LexisNexis for lawyers and MEDLINE for medical practitioners and researchers, have made successful businesses out of providing comprehensive access to published articles and papers.

Similarly, the Los Alamos e-print arXiv has successfully shucked the atoms from physics and mathematics papers—making them instantly available, via the Web, almost anywhere.[8] Initiated in 1991 by Paul Ginsparg, it quickly became indispensable to physics researchers throughout the world. Ten years later it was attracting more than 30,000 electronic paper submissions each year. Many researchers had formed the habit of consulting it each morning for the latest postings in their areas of research. It had dramatically quickened the rhythms of communication and collaboration within the international physics and mathematics communities and had gone a long way toward accomplishing its founder's goal of providing "a level playing field for researchers at different academic levels and different geographic locations."[9] You no longer had to be at an established, recognized center like MIT or Princeton to be an active, effective intellectual participant. Suddenly, geographic location wasn't what counted; what *really* mattered was ready access to a site in cyberspace.

Two more recent online scholarly communities, CogNet (for cognitive scientists)[10] and ArchNet (for architects throughout the developing world),[11] have gone even further. As with arXiv, large online archives are their central attractions. But, like ancient Alexandria, they also provide housing and public spaces for the scholars they attract—not in bricks-and-mortar form, of course, but in the forms of online private workspaces, member profiles, forums, news pages, calendars, job listings, and collaborative workspaces. And as such augmented archives have grown in importance, they have increasingly been viewed as crucial, long-term community resources.[12]

By the year 2000, the writing was on the virtual wall. A National Academies report on the Library of Congress bluntly asserted that digital information was "at the center of contemporary discourse." It continued: "That role is a simple fact, unrelated to whatever e-zealots or bibliophiles might wish to be the case."[13] As evidence, they could point out that popular Web search engines like Google were, by that point, indexing more than a billion online pages.[14] Most of what you needed was just one search away, and most of the rest was at two or three degrees of separation. And with a wireless laptop, you could potentially download any of it anywhere.

FOOTLOOSE CODE

Speech acts (that is, commands, promises, requests, and other utterances with consequences) have also operated in new ways as text has dematerialized and mobilized. Before mobile, written text, speech acts were local verbal events; you might yell a battlefield command or execute an agreement with a word and a handshake, and these sorts of face-to-face acts still carry a great deal of social and legal force. With paper came written instructions from remote commanders, contracts in the mail for signature, formal requests and demands, and the distinction between barristers and solicitors. Now, with electronic text, you can send a command by email, execute an agreement by filling out an online form, or do a drug deal with a pager. Some less scrupulous Muslim men have even figured out that they can utter *talaq* (end a marriage, under Islamic law, by declaring "I divorce you" three times) using the short message services of cellphones.[15]

A line of computer code is an important special case of a written command—one that is intended for execution by a machine. (Today, the best way to delegate a task—and to be sure that it is done right—is to code it for machine execution.) Such code has proliferated explosively in the few decades that we have known electronic computation, and the practical consequences of its growing mobilization have been profound.

In the 1960s punch cards were a standard storage and input medium for computer software; a program and its data might consist of several thousand pieces of perforated cardboard. Their information density was extraordinarily low—much less that that of a printed page. They were handled by means of large, noisy, unreliable mechanical devices: keypunches, sorters, and readers. They were vulnerable to jamming, tearing, burning, and getting left out in the rain. When you threw them out (as you did frequently) you generated a significant quantity of trash. If you were a programmer, you probably spent a lot of time in the keypunch room of a computer center, when you were not walking back and forth to the job submission window with boxes and printouts under your arm.

As magnetic media became cheaper and denser, they gradually took over the role of program and data storage. Reel-to-reel tapes, tape cartridges of various kinds, and floppy disks gradually made code lighter, less bulky, and more portable. When the personal computer arrived in the 1980s, PC software was distributed in shrinkwrapped boxes, and sneakernet (walking with a floppy in your hand) served for transfer among machines. Some early personal computers had no hard drives and only became useful when you stuck in a disk.

Then inexpensive, capacious hard disks, combined with increasingly efficient networking, precipitated a move to online software distribution. Instead of purchasing software from a retail store and lugging it home in a cardboard box, you could now simply download it over the Internet. (This proved particularly convenient for software updates.) The need for floppy disks and CDs greatly diminished, and readers for removable media were no longer essential components of desktop and laptop computers.

Responding to its new environment, code itself evolved. Early programming languages, such as Fortran, had been designed with punch cards explicitly in mind; cards had eighty columns, so Fortran code used lengthy sequences of statements comprising eighty or fewer characters. But network-era languages, such as C++, provided much greater formatting flexibility and allowed programmers to organize code into modular, reusable units known as objects. Then Java fully mobilized code by making it easy to download and execute such units on any networked device—a process that quickly became familiar to browser users, who increasingly found themselves downloading Java applets to run on-screen animations and perform specialized tasks.

Today, code moves fluidly around networks—both wired and wireless—and may take up residence wherever there is available memory. Once resident on a device, it takes control of that device's actions—usefully, in the case of the software that runs your MP3 player, cellphone, or automobile, but destructively in the case of the virus that attaches itself to an incoming email message, takes over your PC, displays a mocking message, then wipes out your hard disk. Function is fungible; you download the reader with the file, the TV set with the television signal.

Dematerialized and radically mobilized code has become allied with the memory and computation capability now embedded in gadgets and machines of all kinds and with the increasingly comprehensive networking of these devices. Together they create a powerful, ubiquitous, rapidly growing structure of commands, conventions, and controls. Increasingly, we live out our daily lives within its invisible boundaries. As the Y2K scare and related efforts to track down and fix errant code revealed, its grip has quickly become global and inescapable.[16]

Code is mobile. Code is everywhere. And code—for both machines and the people who interact with them—is the law.[17]

WEIGHTLESS PICTURES

Primitive pictures came attached to rocks. The earliest extant examples are daubed on the walls of caves, and they are inseparable from the sites of their creation. But, like texts, pictures have lost weight and mobilized themselves in more recent times.

The first move was to the walls of buildings. These provided more convenient working surfaces and allowed the development of sophisticated, precisely controllable graphic techniques—that of fresco painting, for example. The permanent presence of a picture often determined space use. An altarpiece in a church, for example, oriented worship; today it might provide churches an opportunity to charge tourists for admission.

Application of drawing and painting skills to smaller, lighter, more portable surfaces—wooden panels, stretched canvases, sheets of paper, and the like—was an obvious next step. In nomadic and partially nomadic cultures, such as that of the Mughuls, the process of lightening was carried to its logical extreme, and the form of the highly portable miniature emerged. All this had surprisingly far-reaching consequences. It shifted picture production from construction sites to specialized painters' studios, and it greatly facilitated the buying and selling, transfer, collection and accumulation of pictures. And pictures became temporary installations in spaces (particularly galleries) rather than permanent features—allowing for reprogramming of spaces by changing the stuff on the walls.

Portability also allowed topographic views to be made on the spot, often in the open air, then carried away to serve as visual evidence of distant places. During the great ages of exploration, in particular, expeditions were rarely without their skilled artists. The idea of eyewitness visual reportage emerged, and the construction of a vast, continually expanding visual record of the world began.

The trouble with this record, of course, was its vulnerability to inaccuracy, error, and falsehood. But technology soon came to the rescue. Development of perspective projection techniques helped artists (when they wanted to) to achieve higher levels of optical accuracy. Automated projection devices, such as the camera obscura, were even more effective; it seems likely that Vermeer, for one, relied heavily upon them.[18] Fox Talbot took the crucial step of mounting a small sheet of photosensitive material, within a light-tight box, behind a lens. This provided a fast, precise, automatic way of attaching shaded perspective images to silver atoms smeared on glass or paper.

The technological history of photography, from that point on, is a chronicle of miniaturization, acceleration, and increasingly efficient distribution. Emulsions became finer-grained and faster, exposures became effectively instantaneous, large-format glass plates gave way to 35 mm film, tripods mostly disappeared, and cameras became smaller, lighter, and more portable. Photographs accumulated in albums and archives. Through halftoning, photography was combined with printing, and illustrated books, magazines, and newspapers began to flood the world. Photographers, photojournalists, and photo editors increasingly took over creation of the rapidly expanding visual record—a record that gained authority not only from its verisimilitude, but also from the fact that rectangles of emulsion-coated material had, potentially verifiably, been exposed at particular moments in particular places. The photographer, unlike the muralist or studio-bound painter, could claim, "I was there."

The era of photography—of relatively lightweight, easily produced, inexpensive, efficiently reproduced and distributed, optically accurate, verifiable images—lasted for about a century and a half. Then in the 1990s, attached atoms got the coup de grâce. CCD arrays began to replace photographic film on the picture planes of cameras, images were captured and stored as completely dematerialized digital files, and digital photography rapidly became commonplace. Image processing software running on personal computers replaced developing tanks and enlargers and shifted the sites of image crafting from darkrooms to desktops. Digital cameras were combined with cellphones, so that digital images became part of synchronous discourse. Simultaneously, the explosively growing World Wide Web provided a way to distribute digital images with unprecedented speed, cost-effectiveness, and geographic coverage. Pornographers were, of course, among the first to discover that graphic Web sites beat the printed page.

FILMLESS MOVIES

Moving images were almost weightless from the beginning. Some early display devices featured photographs attached to revolving disks or drums, but these were hopelessly cumbersome and only good for a few seconds of flickering motion. Before long, sprocketed transparent film had become the standard—allowing splicing, editing, and movie narratives of unlimited length, the results being projected up to mural size. But unlike murals, of course, these projected images were not attached to the underlying surfaces; they were ephemeral, completely immaterial overlays—repeatable versions of the moving images that had been projected by the camera obscura.

Electronics eventually eliminated even the residual materiality of the projected image. Video cameras encoded moving images as electromagnetic signals, which could be broadcast through the air, carried through a wire, or recorded on magnetic tape, then displayed (for as long as the signal lasted) on a phosphor-coated screen. In the digital era, CCD arrays were quickly incorporated in video cameras, and digital signals, recording media, and displays replaced their analog predecessors.

In microcosm, the technology of classroom and business presentations has recapitulated this chronicle of progressive dematerialization. First there were slates, blackboards, and whiteboards, which presented text and images as thin, removable layers of atoms on flat surfaces. Then came slide and transparency projectors, which shrunk these layers, and displaced them from the display surface to a plane near the light source. And finally, there were PCs, Powerpoint, MPG clips, and video projectors.

By now, cyberspace contains the largest and most mobile accumulation of still and motion pictures the world has ever known—far more extensive than the greatest picture gallery, photo or film archive, or library of illustrated books. Some of it resides in organized image archives, such as Corbis—a direct successor to the stock photo archives of earlier days, but much of it is simply scattered across the Web's innumerable pages and reveals itself piecemeal as we surf. As a result, the average weight of the images we encounter each day is asymptotically approaching zero.

INVISIBLE MONEY

Coins are a fascinating special case of text and images with atoms attached—lots of them. Originally, the value of a coin was that of its constituent metal; minting merely served to standardize the weight and therefore the value. So the forger's only option was to substitute some less precious atoms.[19]

The process of abstraction, miniaturization, and dematerialization of money began when coins became metonyms—fragments representing larger stocks of precious metal or other valuable items stored in some safe place, and available to pay debts if necessary. Size and weight were no longer direct indicators of value; these properties became graphic design decisions. You needed to know *who* was backing up the value with their treasury, so (following Alexander the

Great) the embossed head of the sovereign usually appeared. This allowed currency units to be smaller, lighter, and easier to carry. And, to support geographically extended systems of exchange, they could be moved around at stagecoach or railroad speed.

It was then a short step from metal to paper—that is, to bills of exchange, checks, and banknotes. Paper was lighter still, and particularly convenient for large denominations, which would otherwise require burdensome quantities of coinage. (Smugglers and money launderers are particularly fond of high-value banknotes, like the 500 Euro, since their great value-to-weight ratio allows easy concealment for transfer through checkpoints.) And paper was essentially valueless in itself, so forgers turned their attention from adulterating the material to creating convincing facsimiles of the printed text and images. Once metallurgists, they now became graphic artists.

Paper money further loosened the connection between material content and value, so that value now became an abstract policy variable. By issuing more paper, and postponing redemption, governments could pay for things. Until 1973, the Bretton Woods system attempted to stabilize currency values by defining them in terms of gold, but that practice has now been almost universally abandoned. These days, paper money primarily provides convenience, and serves to reduce transaction costs in the everyday exchanges that drive market economies.

Meanwhile, the telegraph introduced the possibility of wiring money, and created a profitable business for Western Union. By this point, financial transfers no longer required the exchange of metal or paper, but could be accomplished simply by rewriting numbers from column to column in the ledgers of financial institutions. And this rewriting could be triggered by suitable electronic messages. This increased the speed and further reduced the cost of transactions at a distance. The modalities of theft correspondingly multiplied; you could stick up banks or armored cars to get your hands on physical tokens, you could forge those tokens, you could take the white-collar option of fiddling the ledgers to your advantage, or you could subvert the telecommunications system to commit wire fraud.

With computers and digital telecommunications, the ledgers became online databases, software began to manage electronic transfers, and money acquired a complex, hybrid representation.[20] Heavy, durable coins still have a residual function; people carry them in small quantities, and use them to feed parking meters, Coke machines, subway turnstiles, and other electromechanical devices. Banknotes are folded in wallets and are employed for larger purchases. Light plastic credit cards and debit cards are also carried in wallets; these carry electronic identification and can be inserted into reading devices that will trigger electronic transfers or disgorge banknotes. Transponder-equipped smart cards and keychain devices (such as Speedpass and SmarTrip) transmit billing information to gas pumps, subway turnstiles, vending machines, and other payment points.[21] Online, you don't even need the plastic; the credit card number suffices to initiate a transaction. Some pioneering dot-coms, like PayPal, have gone further, demonstrating the possibility of emailing money or even beaming it from one portable, wireless device to another. In larger quantities, completely abstracted and dematerialized money now flows from institution to institution at breathtaking speed through electronic trading systems and funds transfer networks.[22]

Encrypted electronic transfer can cut transaction costs to the point where micropayments become imaginable. Potentially, this radically reduces the granularity of transactions. Instead of purchasing a complete book, you might make a micropayment each time you access a page online. Similarly, you might make a micropayment whenever you download and replay a recorded musical performance. Conversely, you can make huge payments—the equivalents of truckloads of banknotes—in an instant. Electronics releases us from traditional constraints on transfer scale and granularity.

Furthermore, financial transactions need not be restricted to the direct digital equivalents of coins and banknotes; they can involve complex abstractions, such as derivatives—entities that could not exist anyplace other than cyberspace. And transactions may no longer be under direct human control—the electronic equivalents of handing over cash; they may become highly automated, as with programmed stock trading.

Architecture has reflected money's evolving abstraction, dematerialization, and retreat from visibility. Banks used to be built around their fortified vaults—the places where wealth accumulated in the most literal sense. They were serviced by highly visible armored vehicles and located prominently within communities to facilitate deposits and withdrawals of coins and notes. Later, automated teller machines combined electronic transfer capabilities with miniature, automatically operated vaults to provide more effective geographic coverage and 24-hour service. At around the same time, retail stores and restaurants installed unobtrusive electronic card readers at points of sale in place of grand mechanical cash registers. Then networked personal computers allowed financial transactions to be performed online, just about anywhere. As money has emancipated itself from the last vestiges of materiality, sites of accumulation and transfer—once prominent, architecturally celebrated urban elements—have waned into commodified electronic boxes, anonymous server farms, and out-of-sight back offices.

MOBILIZED MUSIC

To mobilize music, you need to detach it not only from the site of its initial performance but also from the *moment*. You have to displace and time-shift vibrations in the air.

Traveling troubadours did so by memorizing tunes and traveling from place to place. Musical themes and ideas diffused through face-to-face contact at performance times. Maybe you improvised together—much as jazz musicians still do. You had to be there.

From the ninth century in Europe, musical notation has provided a way to externalize the musician's memory—to write down and preserve specifications for performances of a work. The first music books then functioned, as Nelson Goodman put it in a famous definition, as "the authoritative identification of a work from performance to performance." This served, among other things, to establish a uniform liturgical practice dictated by the church authorities, and to limit the scope for improvisation.

The practice of writing down musical scores distinguished logically between composers and performers (though, of course, one person might sometimes perform both roles), and separated music, in subtle but profound ways, from art forms such as painting. Painters do not normally set out to comply with scripts or scores; their works are the unique products of particular authors, working at particular times, in particular places. In Goodman's terminology, they are *autographic*. But scored or scripted works may have indefinite numbers of performances, at different moments and locations; they are *allographic*.[23]

Allographic works may be performed not only by human artists, but also by mechanical devices equipped with the capacity to follow appropriately encoded scores. At the end of the nineteenth century, Edwin S. Votey's pianola vividly demonstrated this possibility. In its earliest form, it consisted of a large wooden cabinet that stood in front of a standard piano. Felt covered wooden levers, protruding from the rear, were aligned with the piano keyboard. Sequences of notes were encoded as perforations on paper rolls, which controlled the operation of these mechanical fingers. Pianola rolls could be recorded at a keyboard, or more usually, perforated by a technician working directly from a score.

In the decades that followed, more sophisticated player pianos and reproducing pianos emerged, and supported two distinct streams of practice. Frequently, player piano technology was employed to capture and reproduce keyboard performances by famous artists—producing, for example, an important legacy of jazz and ragtime from that era. At other times, composers used the technology as a way to transcend the limitations of the human hand. In the 1930s, the era of the player piano was brought to an end by the depression and increasing competition from the less expensive, more compact and versatile gramophone.

The gramophone, like its immediate ancestor, the phonograph, dealt directly in vibrations. In 1877 Thomas Edison had successfully demonstrated a "tinfoil phonograph" based upon a foil-wrapped cylindrical drum mounted on a threaded axle. A diaphragm connected to a stylus picked up acoustic vibrations and etched corresponding wave patterns on the rotating foil. For playback, a pickup stylus retraced the wave patterns and transmitted the vibrations to the more sensitive diaphragm of the "reproducer." Edison recited, "Mary Had a Little Lamb" into the mouthpiece and replayed the mechanical, tinny-sounding echo.

There was a compelling symmetry to the Edison process, and Edison saw his device as a personal recorder—much like later tape recorders. But by the 1890s a commercial recording industry was developing. Musicians recorded performances, then record companies mass-produced and distributed them as closed, branded items of intellectual property. Eventually there emerged a commercial and legal regime under which consumers could do little more than replay recordings in private; their rights to public replay, to re-recording, and to incorporation of recorded material in new works were severely restricted. Recording interests deployed a criminalizing rhetoric of "theft" and "piracy" to protect their position.

As the popularity of radio broadcasting rose rapidly in the 1920s, the recording and broadcasting industries converged both technologically and commercially. The emergence of RCA Victor—formed after the Radio Corporation of America's acquisition of the Victor Talking Machine Company—was a decisive moment. Electric microphones and amplifiers replaced the phonograph's mechanical systems. Recording and broadcast were both centrally concerned with capturing, recording, and distributing electrical audio signals. Sound quality improved, artists could have individual microphones instead of clustering around shared recording horns, and recordings could be edited and recombined.

From the birth of the phonograph to the dawn of the digital era, competing recording media proliferated—evolving in the direction of greater fidelity, reduced cost, and reduction in bulk and weight. Phonograph cylinders were manufactured from tinfoil, then wax-coated cardboard, then solid wax. Gramophone disks, which first emerged in the 1890s, were made from hard rubber, shellac, vinyl, and even, as a curiosity, from chocolate, with various diameters, and speeds of 78 rpm, $33\frac{1}{3}$ rpm, or 45 rpm. Wire recorders enjoyed brief popularity. Magnetic tape was packaged in reel-to-reel and cassette formats, in varying widths, and with differing numbers of tracks. It was an extended effort to minimize the attachment of atoms to music. By 1979, with the introduction of Sony's Walkman, records and players had shrunk to wearable size.

The inevitable move from analog to digital recording formats began in the 1980s, with the introduction of compact disks (CDs) and digital audio tapes (DATs). But the personal computer and the Internet introduced a far more revolutionary possibility—that of capturing, editing, and replaying audio files at any computer, storing them online, and distributing them throughout the world, at high speed, in dematerialized form. Standards to support this new practice began to emerge, and eventually MP3 took hold globally.[24] To the consternation of the Recording Industry Association of America, there was soon software to rip MP3 files from CDs and the digital equivalents of the Walkman (notably the Rio MP3 player) for storing and replaying them.

Furthermore, an infrastructure to support online music distribution—particularly directories of available files, search engines, download managers, and software MP3 players—quickly followed. Napster, which appeared in 1999, was the most radically and visibly challenging to the established distribution systems. The key idea of Napster was to coordinate a highly decentralized, peer-to-peer, Web-based distribution system.[25] It maintained a central, online directory of available MP3 files but did not store the files themselves on its servers. Instead, Napster users stored files locally and made them available for sharing. When users wanted files they did not have, Napster's search engine located these files and downloaded them. As the new century dawned, Napster's intellectual (if not commercial) success could be measured by the way that it was clogging networks and attracting lawsuits from the established recording industry. It had completed the job of dematerializing and radically mobilizing recorded music.[26]

GRANARIES TO SERVER FARMS

In the B.D. (Before Dematerialization) era, settlements were built around fixed, central sites of material accumulation—typically of excess, storable agricultural produce such as grain. Consider Palladio's villas in the Veneto, from about the fourth century B.D., for example. They were surrounded by grain fields, vineyards, orchards, and housing for barnyard animals. The harvested grain was stored in the attic (where it also provided insulation), the wine and cheese were stored in the basement (where they remained at an appropriate temperature), and the consumers lived on the grand *piano nobile* sandwiched in between. Since transportation relied upon muscle power, it mattered a great deal that distances from the fields to the storage areas were short, and that those from storage to consumption were even shorter. There was a clear spatial logic and a wonderful visual clarity to it all; you could take it in at a glance.

Then, as increasingly efficient transportation technologies emerged, large cities were organized around extensive distribution networks that incorporated diverse, specialized production, accumulation, and consumption sites. Since transportation costs remained fairly high relative to the value of the things transported, and since transportation over long distances took significant amounts of time, the locations of these sites were largely determined by accessibility considerations; warehouses, for example, needed to be sufficiently accessible both to the factories that supplied them and to the customers they served. Thus, the great industrial cities of the second century B.D., such as Chicago, were shaped by their rail networks. And those of the first century B.D., like Los Angeles, were given form by their freeways. To an observer on the ground, these extended systems made little sense, but from high in the air you could immediately see how they worked.

At the dawn of the A.D. (After Dematerialization) era—around Y2K in the older calendar system—yet another pattern emerged. Servers and server farms embedded in high-speed telecommunication networks became the crucial, characteristic accumulation sites of new urban formations. Unlike their predecessors, such as granaries, warehouses, bank vaults, and library book stacks, these sites stored dematerialized assets—that is, texts, images, video, music, computer code, and money—in digital format. They were notably inconspicuous, since they consumed relatively little space, were kept discreetly anonymous for security purposes, and offered little opportunity for architectural celebration. Furthermore, they were spatially ambiguous; the common practices of creating backup copies at multiple locations, caching copies at locations near to likely users, distributing databases over numerous different servers, and continually migrating material to new storage hardware made it difficult to specify exactly where particular items were located. (A Web document is not, for example, like a valuable manuscript, such as the Book of Kells, which is permanently stored at a well-known, specific location in Trinity College, Dublin.) And as networks became increasingly pervasive and efficient, the costs of transporting digital items to and from servers were reduced to insignificance relative to the value of the goods; centrality of location no longer mattered much, and important servers could operate effectively in out-of-the-way locations.

Finally, with the development of inexpensive, miniaturized electronics, the production and consumption sites related to these servers increasingly dispersed, mobilized, and began to associate themselves with footloose individuals rather than fixed architectural settings. Today, you can download to your portable wireless devices, you can upload from them, and it is of little importance where the servers you access happen to be. Mostly, you don't know or care. The more you deal in dematerialized goods, the less location and distance concern you. And the less visible are the relationships that really matter.

6

DIGITAL DOUBLIN'

In *Ulysses*, James Joyce taught us to see a city in a new way. From the moment that "stately, plump Buck Mulligan" wakes to lather and shave, around eight o'clock on the morning of 16 June 1904, until the very second that Molly Bloom's memory ambiguously replays "yes I will Yes," Dublin appears as the site of multiple threads of consciousness unspooling in parallel while the inhabitants go about their business.[1] And these threads are continually being interwoven—revealing relationships and disclosing an intricately intersecting structure of motives and desires—as the protagonists encounter and reencounter one another.

This fabric of simultaneous, cross-connected subjectivities is constructed mostly through walking, taking the tramcars, and finding face-to-face encounters in the varied public spaces and private chambers of the city.[2] Letters and telegrams arrive, and there is an occasional word to the telephone, but Bloomsday unfurls at the very dawn of the wireless era; there is no FM 104 drive-time radio, and there are no cellphones. Contiguity enables connection, distance means disconnection. When Leopold Bloom departs for Dlugacz's to purchase a kidney for breakfast, he leaves Molly—crucially—with her uninterrupted private thoughts. If he had called her from the porkbutchers—maybe asking her to take off the kettle he had left boiling and scald the teapot—it might all have turned out differently.

At one point, however, Bloom does allow himself to speculate about "a private wireless telegraph which would transmit by dot and dash system the result of a national equine handicap (flat or steeple-chase) of 1 or more miles and furlongs won by an outsider at odds of 50 to 1 at 3 hr. 8 m. p.m. at Ascot (Greenwich time) the message being received and available for betting purposes in Dublin at 2.59 p.m. (Dunsink time)."[3] A century later, this embryonic possibility has puffed up into a pervasive post-Joycean reality, and, as Dubliners now gab endlessly into their private wireless telephones, they construct their interlocking narratives in a voice that Bloom would hardly have heard—the electronic present continuous.

ELECTRONIC PRESENT CONTINUOUS

This characteristic voice of our time has gradually emerged as telecom-munications systems have grown in capabilities and coverage. Tele-graph and wireless telegraph operators employed it as they tapped out SOS messages describing perilous predicaments. In Australia, from 1888 onward, the telegraph lines were cleared for the annual running of the Melbourne Cup, so that news of the winner could be flashed around the continent in near real time. (Consequently, the whole country still comes to a halt while the race—now broadcast on radio and television—is run.) And at the opening of the first Australian federal parliament, in May 1901, the telegraph was used to create a nationwide "imagined community."[4]

The telephone, by its very live-spoken, circuit-switched nature, provided a natural medium for the electronic present continuous. And it was soon followed by live radio broadcasts of performances, sport-ing events, and breaking news. By the 1930s, in *Finnegans Wake*, Joyce could imagine a radio—a "tolvtubular daildialler, as modern as tomor-row afternoon and in appearance up to the minute, . . . with a vital-tone speaker" sucking a whole, complex, unfolding world into H. C. Earwicker's ear. It was, we are informed, "capable of capturing sky-buddies, harbour craft emittences, key clickings, vaticum cleaners, due to woman formed mobile or man made static and bawling the whole hamshack and wobble down an eliminium sounds pound so as to serve him up a melegoturny marygoraumd, eclectrically filtered for allirish earths and ohmes. . . ." HCE—Here Comes Everybody, yes, and Here Comes Electronics, too.

As the British Empire slowly expired, the red-colored areas of the map preserved a tenuous sense of community through live, short-wave broadcasts of international cricket matches—cutting across time zones and keeping listeners awake at odd hours of the night. In the Soviet Union, wired broadcasts—first to public loudspeakers in the Stalin era, then to private Radio Mayak and Radio Rossiya boxes in the Brezhnev era—provided the government with a direct line to the people, plus a way to keep their attention by broadcasting concerts and hockey games.[5] Across fewer time zones, national baseball, football, basketball, and hockey leagues and broadcasts performed a similar unifying function in the United States. Baseball's most famous moment was given much of its enduring resonance by Russ Hodges's widely broadcast play-by-play—the electronic present continuous pegging the needle of rhetorical hyperintensity:

> Hartung down the line at third, not taking any chances, Lockman without too big a lead at second—but he'll be running like the wind if Thomson hits one. Branca throws . . . There's a long drive! It's going to be, I believe . . . The Giants win the pennant! The Giants win the pennant! The Giants win the pennant! The Giants win the pennant! Bobby Thomson hits into the lower deck of the left field stands! The Giants win the pennant! And they're going crazy! They're going crazy! Oh, ho!"[6]

EYEWITNESS NARRATIVES

By eight o'clock in the evening, on Sunday, 30 October 1938, when Orson Welles began presenting *The War of the Worlds* over the Columbia Broadcasting System, the nuances of the electronic present continuous were long familiar. Listeners knew how to interpret them—and that provided an unprecedented dramatic opportunity. Of course it was just a scary radio play about an unexpected alien attack from the skies, with the characters breaking in to "regular programming" to provide "live eyewitness" accounts of fiery destruction in New Jersey and New York. The actors addressed the audience directly, describing events supposedly unfolding before their very eyes:

Ladies and gentlemen, this is the most terrifying thing I have ever witnessed . . . Wait a minute! Someone's crawling out of the hollow top. Someone or . . . something. I can see peering out of that black hole two luminous disks . . . are they eyes? It might be a face. It might be . . .

But, as the *New York Times* reported (in the past tense) next morning:

Despite the fantastic nature of the reported "occurrences," the program, coming after the recent war scare in Europe and a period in which the radio frequently had interrupted regularly scheduled programs to report developments in the Czechoslovak situation, caused fright and panic throughout the area of the broadcast.

Telephone lines were tied up with calls from listeners or persons who had heard of the broadcasts. Many sought first to verify the reports. But large numbers, obviously in a state of terror, asked how they could follow the broadcast's advice and flee the city, whether they would be safer in the "gas raid" in the cellar or on the roof, how they could safeguard their children, and many of the questions which had been worrying residents of London and Paris during the tense days before the Munich agreement.[7]

It did not matter that the events being described were imaginary—and, indeed, wildly implausible. There was no simple way to know that the "explosions" were special effects. If you were too far away to be a potential eyewitness yourself (it was clever of Welles to set ground zero on an isolated farm in New Jersey) and too uneducated to catch the scientific howlers (Welles deployed a "great astronomer" and other scientific authority figures to deflect critical scrutiny of the details), it was impossible to distinguish fact from falsehood. Electronic signals had demonstrated that they could construct the most out-landish of beliefs and provoke people to act instantly on those beliefs.

It was the power of the electronic present continuous voice—the presumed direct connection to simultaneously unfolding, distant events—that caused the trouble. The damage had been done long before Welles got to his final (retrospective voice) disclaimer:

This is Orson Welles, ladies and gentlemen, out of character to assure you that *The War of the Worlds* has no further significance than the holiday offering it was intended to be—the Mercury Theater's own radio version of dressing up in a sheet and jumping out of a bush and saying "Boo!" Starting now, we couldn't soap all your windows and steal all your garden gates by tomorrow night—so we did the best next thing. We annihilated the world before your very ears, and utterly destroyed the CBS. You will be relieved, I hope, to learn that we didn't mean it, and that both institutions are still open for business . . .

Today, live eyewitness accounts can be transmitted not just by expensively equipped broadcasters, but also by anyone with a cellphone or instant messaging system. Husbands can describe to their wives the choices they see before them on the supermarket shelves, teenagers can tell their friends where they are and who is hanging out with them, street demonstrators can point out the current positions of the police to their comrades, and celebrity spotters can track the movements of George Clooney through the produce section at Balducci's. When the hijacked jetliners hit the World Trade Center towers, around nine o'clock in the morning on Tuesday, September 11, 2001, the unfolding events were narrated in thousands of phone calls and instant messages; it was a desperately improvised, scaled-up replay of *War of the Worlds*—but with nobody to step reassuringly out of character at the end.

In cities today, electronically propagated narratives flow constantly and increasingly densely. These narratives—superimposed, as they are, on real space in real time—act as feedback loops recursively transforming the very situations that produce them. And, like all narratives, they are of ambiguous reliability—constructed from facts, fictions, and falsehoods in whatever sorts of mixes and combinations their authors care to contrive. In Hugh Kenner's well-known reading, *Ulysses* is the voice of a cybernetic mechanism—"a huge and intricate machine, clanking and whirring for eighteen hours";[8] now narrative has rebounded from the pages where novelists had consigned it and has itself *become* the city's cybernetic machinery.

In the late 1920s (not long before the publication of *Ulysses*), Dziga Vertov, in *Man with the Movie Camera*, constructed a closely comparable, day-long urban portrait—this time of a Soviet city.[9] Like the novel, the film begins with the protagonists awakening in the morning and carries the narrative through the day's activities to the late evening, intercutting multiple strands of action as it goes. It self-consciously introduces, on-camera, the camera itself, the cameraman, the film, the projector, the projectionist, the theater, the audience, the orchestra, and the flickering screen. And, every now and again, it tweaks us into sudden awareness of the paradoxical nature of its enterprise—for example, by pulling back to reveal that what we took to be a direct view of the city is actually a filmed view, being projected at a later moment before the theater audience. It reminds us that the movie camera and theater represent a dismemberment of the old camera obscura—shifting the glowing projection screen to a different darkened box from the camera lens and temporally displacing the beam of light that enters the system from that which exits.

Vertov's ribbon of spliced celluloid also foregrounds the central, inescapable contradiction of all recorded narrative—the bafflement that, for example, animates *Tristram Shandy*, in which Tristram's Uncle Toby starts an autobiography that is slower-paced than the events it describes and falls more and more hopelessly behind as life inexorably goes on. You simply cannot sync recorded narratives with the events they describe. That's what you need tense for. You can set your narrative in the past, so that it continuously recedes into the distance. You can set it at some point in the future, and describe what will be, so that real time eventually catches up with your predictions and passes them by. Or you can try the eyewitness present continuous voice— but you had better be quicker than Uncle Toby. In any case, the spans of time occupied by the actual events unfolding, the writer writing about them, and the reader reading, will generally be different.

But link a video camera to the Web, and these distinctions collapse; Uncle Toby's paradox suddenly dissolves. You now get *Ulysses II*, unfolding second by second, in real time. Today you can surf into the Web site of the *Irish Times*, from anywhere in the world, at any time, to get live Webcam views of the O'Connell Bridge and the Liffey.

Or you can click to the Dublin Corporation's numerous traffic cameras for views of the streets—now car-choked—that Leopold Bloom, Stephen Dedalus, and Buck Mulligan strode. And I'm not even counting the private surveillance cameras scattered discreetly around Temple Bar.

It's not just Dublin, of course. A century after Buck Mulligan roused himself to peer down the stairs, the multiple narrative perspectives of the characters in *Ulysses* have become endlessly proliferating, live video feeds from the world's cities. In the emerging Webcam era, the city's cybernetic loops are not only audio and text narratives, but video streams as well. As cities electronically reflect back on themselves, telling is now joined by showing.

Like Joyce in far-off Trieste, distant Web surfers can construct minutely detailed accounts of the day's doings in places that may interest them—not by a novelist's occupation of the minds of his characters, nor by a filmmaker's shooting and editing, but by following hyperlinks to switch among available video viewpoints. And as to the mode of consciousness induced by this strategy, it is precisely that detected by Hugh Kenner in *Ulysses*—the mode of a "sardonic, impersonal recorder, that constantly glints its photoelectric eyes from behind the chronicle of Bloomsday."[10]

MANN WITH THE VIDEO CAM

Joyce disclosed Dublin by mobilizing his characters, and film or video makers do so by mobilizing their cameras. Thus the agility of the available cameras matters. Indeed, much of the now-standard language of film derives directly from the bulk and clumsiness of early movie cameras.

Vertov's camera, as the film itself reveals, was a large, hand-cranked box on a tripod. (Uncle Toby would have appreciated that, the faster the cameraman cranked, the slower the eventual projected scene unfolded.) Consequently, many of the shots in *Man with the Movie Camera* are framed statically; there are a few pans, and only an occasional, daring take from the back of a moving truck. The film's sense of frantic movement through the city is mostly achieved through skillful construction of fast-paced montages.

Hollywood has long attempted to overcome the limitations of the camera, and to achieve greater fluidity, by introducing sophisticated motion machinery. For example, the famous extended tracking shot through a Mexican border town, at the beginning of Orson Welles's *Touch of Evil*, was a tour de force of the mechanized camera crane. And, more recently, robotically controlled cameras have been increasingly used for precisely choreographed special effects work.

The miniaturization of film and video technology has also loosened things up. In particular, the development of light, handheld film and video cameras has enabled a first-person, body-based style; the camera operator's mobile eye seems to become the viewer's eye, and the voiceover commentary seems a whisper in the viewer's ear. This is characteristic of home videos, and it has been common in documentaries. It has also been used to great effect in a few theatrical films, such as *The Blair Witch Project*. It was a short technological step from this to the wearable, head-mounted video camera—a possibility I first saw instantiated, in the early 1990s, when a talented Media Lab student named Steve Mann walked into my MIT office with video rolling.[11]

If you have enough wireless bandwidth, you can connect a handheld or head-mounted video camera to a transmitter. You can (as Mann did) transmit sequences of images to a Web site for later viewing. Or, as the rollout of G3 cellphone technology demonstrated in 2001, you can transmit live video through a wireless telephone. Thus the "cinema-eye" and the "radio-ear"—technologies that Vertov had seen as alternative ways of reaching the masses—are combined and mobilized.

The ultimate in camera miniaturization and mobilization is the "virtual camera" of three-dimensional computer graphics, which shrinks to a weightless point. The effect of this was vividly evident in the early Disney computer animation *Tron*, in which the "camera" performs high-speed gymnastics that would be unthinkable with a physical device. Now, such effects are commonplace in videogames, and in the exploration of online "virtual cities," such as *Virtual Helsinki*. If model-based video eventually replaces today's image-based systems, as some researchers expect, the dimensionless, infinitely mobile camera will become the norm; instead of watching a sporting event through the medium of a few physical cameras and the shots selected by a director, a viewer might "fly" at will through a live three-dimensional model of the stadium.

Through miniaturization and mobilization of video cameras, wireless connection, and model-based video, the stiff, distant images provided by fixed Webcams will soon be supplanted by fluid, eyewitness perspectives. Electronic images of the city will seem less like digital descendents of Vermeer's view of Delft and more like retinal traces from Stephen Dedalus in motion.

ELECTRONIC TWINS

In speech, the shift of reference from present to past events is signaled by tense; "I *am* walking down Grafton Street" becomes "I *was* walking down Grafton Street." But video doesn't have participles. So, there is no perceptible difference in the image when a delay is introduced into the transmission of a video signal; the instant replay of a football play looks exactly the same as the live transmission you saw a moment ago. There is simply an unmarked tense drift.

When the hijacked airliners hit the World Trade Center towers on September 11, the moment was captured on video. For hours afterward, in a kind of pornography of explosive violence, the clips were replayed over and over again. On screen, it was hard to tell the difference between the identical towers, and tense drift added another layer of nightmarish ambiguity. As growing numbers of people, around the world, tuned in to these images on their television and computer screens, it often wasn't clear to them whether they were watching events in real time or replays. As the broadcasts cut back and forth between live and recorded images of the same unfolding scene, this tragedy drifted away from the classical unities.

In Manhattan, apartment dwellers simultaneously saw the smoke framed by their television screens and by their windows. Banks of screens in electronics stores flashed the scene in Warholesque multiples. The LEDs in Times Square brought Downtown to Midtown. In the darkened Postmasters Gallery, Wolfgang Staehle's live-feed video projection of the Lower Manhattan skyline (which had opened on September 6) sliced the unfolding chaos into a sequence of eerily silent, luminous landscape stills.[12] And, in a suddenly sinister way, Staehle's successive stills recalled a Photoshop demo; duplicate, erase, paint in some sky effects, eventually substitute something else.

Today, a fundamentally new urban condition is emerging—one that was anticipated by Joyce's repeated, sardonic reference to Dublin as Doublin', a city marinated in narrative, and inescapably bound up with narrative's capacity both for reflection and for duplicity. Multiplying thousands of electronic eyes and ears continuously capture the city's unfolding, interwoven narrative threads, and spin them out into cyberspace. Some of these threads are ephemeral, and disappear instantly. Others sit on voicemail, email, and other servers for a while, then are deleted or automatically fade away. Yet others accumulate permanently, to form an expanding, long-term electronic memory trace. And continuously, the narratives—drifting in tense and dislocated spatially—leak back out again into the city through proliferating earpieces, eyepieces, speakers, and screens. In countless spatially and temporally displaced, inherently ambiguous, recombinable fragments, cities electronically double themselves.

To find something out, or to get something done in a city, you now have a choice. You can navigate the bricks-and-mortar half in the time-honored way, or increasingly, you can switch to its electronic twin.[13]

7

ELECTRONIC MNEMOTECHNICS

If I send you an email message saying "stop," it will show up at whatever time and place you choose to download, and you probably won't know what to make of it. But if I send you a SMS message it will arrive instantly, you will have a definite temporal context, and you will probably understand that I want you to stop something you are doing right now, *wherever* you may be. (Similarly, the muezzin's call to prayer is temporally specific but spatially universal; if you're faithful, you're meant to heed it right now, wherever you happen to be.)[1] If you encounter a stop sign as you approach a street intersection, there is a definite spatial context to give meaning to the message; you are supposed to stop right there, *whenever* you approach the intersection. And, if a cop with a gun yells "stop" as you flee, the spatial and temporal contexts are both definite; he means right there, right now.

Some sorts of messages, like the old IBM slogan "Think!," are less context-specific. It also does not matter, in one sense, when and where you read the Theorem of Pythagoras; it is universally true. However, of course, it will only rarely be relevant to what you are currently doing.

In general, the meaning and relevance of a message may vary with the spatial and temporal contexts of its reception. And messages can be classified according to the spatial and temporal contexts they require, as illustrated in the following table.

	SPATIALLY INDEFINITE	SPATIALLY SPECIFIC
TEMPORALLY INDEFINITE	Theorem of Pythagoras	Stop sign
TEMPORALLY SPECIFIC	Broadcast sports scores	Air traffic controller's command

This imposes some requirements on message senders and shapes their technology choices. If you want to send a message that is spatially specific but temporally indefinite, you can inscribe it on a wall, and if you want to send one that is specific to a portable object, you can affix a label. Conversely, if you have a sports score or race result that is of instant interest over a wide area, you can broadcast it. If the message is location-specific but varies with time, as with the flight information posted at an airline departure gate, you can use a LED display. If it is spatially and temporally universal, as with a mathematical or scientific result, you might publish it in multiple copies for wide distribution and consultation at indeterminate future dates. If the intended recipient is within sight and earshot, as when you yell to a running child, you do not require any additional technology to assist in achieving spatial and temporal specificity. But if the recipient is out of immediate sensory range—as when an air traffic controller gives instructions to a pilot—some sort of location awareness technology is necessary; specifically, the controller watches the aircraft on a radarscope.

In our wirelessly interconnected world, electronic location awareness is becoming commonplace.[2] Wireless signals, like sounds, encode information about their sources. By various means, it is possible to discover their directions of origin, and deduce their likely distances. Consequently, the coordinates of portable wireless devices, and of the objects and bodies that bear them, can be tracked.[3] Unlike the world of the ancient nomad, the world of the wireless cyborg is inherently one of traceable positions and trajectories.

This provides an increasingly crucial new way to weave together cities, their inhabitants, and digital information. You can tag information according to its geographic origin, you can keep track of relationships between places and bits, and you wirelessly deliver bits to exactly where they are needed and make sense. Instead of being part of the city's fixed infrastructure, for example, a stop sign might automatically appear on your dashboard, in response to current traffic conditions, as you drove up to an intersection.[4]

LOCATION-TRACKING TECHNOLOGIES

One approach to extracting spatial coordinates is simply to measure the strength and direction of a received signal to get an indication of the transmitter's distance from you. This is particularly useful if you are searching for a transmitter, since the signal gets stronger as you move toward it. Thus, in a common application, small transmitters in automobiles allow them to be tracked and located when they are stolen.

If you are using a cellular system, the cellular operator can keep track of the base stations to which you link as you move around. Measurements of signal strength are used to control switching between base stations. When a call is made or received, the system detects the nearest base station and logs a phone location code in a database. In urban areas with high densities of base stations, this typically establishes location to within a few hundred feet, and also allows approximate motion tracking as mobile callers move from cell to cell. As cellular systems move to higher bandwidths and carry heavier traffic, cells will get smaller and this sort of tracking will become more precise.

Another way to get more precise location codes it to measure signal strength at three base stations. This provides sufficient information to compute a location code by triangulation. This will probably be done increasingly as phone location codes are put to more and more practical uses.

Phone location codes are, of course, a decidedly mixed blessing. They provide telecommunications companies with billing information and traffic statistics. If available to governments, they can be used for the most chilling kind of comprehensive citizen surveillance, and if available to telemarketers they can be used to direct pinpoint junk mail right to your phone.[5] They can also be used by the police and the military for real-time tracking of targets; terrorists have learned that cellphone calls may be answered by missiles. More benignly, they allow 911, roadside emergency, and taxi calls to be tagged automatically with caller coordinates. In today's Dublin, you wouldn't need a novelist's omniscience to follow Leopold Bloom, Stephen Dedalus, and Buck Mulligan around the city; you could just track their cellphone usage. And if Leopold could get access to the logs, he could figure out precisely what Molly was up to.

Yet another approach is to set up special radio beacons at known locations; wireless-equipped travelers can then measure the time taken for signals to reach them from different beacons, convert those times to distances, and thus calculate their current positions by triangulation. This is, of course, a translation of ancient surveying and navigational techniques into wireless processes, and there are many variations on the idea. Loran navigation systems, for example, employ high-powered radio beacons at fixed locations. At a smaller scale, some three-dimensional digitizers and location trackers make use of ultrasonic beepers and strip microphones. And ultra-wideband localizers can now make use of radio frequency (RF) transceivers distributed in the environment to provide centimeter accuracy over kilometer distances.[6]

THE GPS SYSTEM

The most comprehensive and sophisticated infrastructure for wireless location and navigation is provided by the global positioning system (GPS), which was completed, in its original form, by the U.S. Department of Defense in 1994.[7] This system derives distances (and hence locations) from the amounts of time taken by radio signals to travel from satellite transmitters to ground-based receivers.[8]

The GPS system is based upon twenty-four satellites that circle the earth in a pattern that allows at least five to be seen from any location at any time. Each satellite contains a very precise clock, and the exact position of each satellite is known at any moment. The satellites transmit signals specifying their clock times and locations, and the times taken by these signals to reach a receiver establish the distances of the satellites from the receiver. Four of these signals provide enough information to compute the three-dimensional location of the receiver.

Remarkably, the system (when it isn't deliberately made less precise) is accurate to within a few meters—good enough to navigate city streets. It was designed initially for military guidance systems but was soon extended to civilian ships and aircraft as well. By the year 2000, GPS systems were becoming routine equipment in automobiles, and inexpensive pocket-sized models were popular equipment for hikers. They were getting built into cellphones to provide more accurate location-based 911 capabilities.[9] And they were even being incorporated into digital cameras, so that images could be stamped not only with the time of exposure, but also the place. By 2002, single-chip GPS systems had been announced—making it possible to integrate GPS capability into devices as small as wristwatches.

INDOOR TRACKING

Unfortunately, architectural features block GPS signals, so GPS tracking does not work indoors. Beacons located throughout a building can provide similar location-tracking capabilities, but the task is complicated by the fact that buildings typically contain metal and other materials that affect the propagation of radio signals, producing interference and dead spots.

The Cricket system, developed at the MIT Laboratory for Computer Science, deals with the difficulty by employing beacons that emit both radio and ultrasonic pulses.[10] Since these travel at different speeds, the difference in their arrival time at a receiver can be used to compute distance—much as you can compute the distance of a lightning flash by counting the seconds before you hear the sound of the thunder. By discovering the nearest beacon in this fashion, a location-sensitive device can determine which room it is in.

In the future, dense grids of inexpensive location beacons will become standard architectural features. These will enable indoor pedestrian navigation in much the same way that GPS supports automobile navigation, and they will allow spaces to advertise the facilities and services that they provide. The experiences of exploring a museum, searching for a product in a department store, or finding your car in a large parking garage will be transformed.

RADAR, TRANSPONDERS, AND RF TAGS

Radar and sonar tracking stations work in still another way—by bouncing directional signals off moving planes, ships, thunderclouds, and other objects of interest. Conversely, mobile radar and sonar systems in land, sea, and air vehicles bounce signals off the surrounding topography to provide navigational information. The principle may be generalized; anything that propagates a signal into its surrounding environment, from a chirping bat to an 802.11 base station, can potentially employ the reflections of that signal to construct some sort of picture of its environment. Paul Virilio, ever alert to such things, has noted the panoptic potential of ubiquitous broadcasting; he imagines the world's television and FM radio transmitters organized into a gigantic surveillance engine that "detects any activity, any movement, legitimate or otherwise."[11]

If you embed a transponder in an object, you can go further, and interrogate the object by directing a radio signal at it. Ground stations interrogate transponders in approaching aircraft to determine whether they are friendly or not, and electronic tollbooths interrogate transponders in automobiles to charge tolls. By broadcasting location queries to transponders in GPS-equipped vehicles, you can keep constant track of where they are and how fast they are traveling— a capability of obvious interest to managers of delivery fleets, rental car operators, transportation planners who want traffic flow data, traffic police who want to keep speeding in check, and three-letter agencies.[12]

Early transponders were bulky and expensive, but it is now possible to produce tiny, inexpensive ones on microchips and even to dispense with batteries or external power sources by harnessing the power of the incoming signal. RF tags, based upon this principle, are increasingly being sewn into clothing, embedded in electronic devices, and implanted in both animals and humans.[13] We are rapidly approaching the point where anything might respond instantly to a wireless query about its identity, current location, or some other property.

A transponder or RF tag might reply with something as simple as an identification number, much like the barcodes that are read by optical scanners. It might directly provide detailed information about its host object, as would a printed label. Less directly, it might provide the URL of a Web site that contained detailed information on its host. And if it is connected to sensors, it can provide current information about its host's condition; for example, the California Air Resources Board has proposed to equip vehicles with transponders connected to their emissions monitoring systems, so that malfunctioning vehicles automatically identify themselves—the electronic equivalent of a scarlet A on your forehead.[14]

The most enthusiastic proponents of RFID project a world in which pretty much everything has a smart tag with a unique, 96-bit electronic product code (EPC) that can be read by a simple radio device and used instantly to connect it to the Internet and access an online database that contains the object's specifications and history—its detailed digital shadow.[15] Unlike printed labels, the electronic shadow is of unlimited size and can be dynamically updated. And by contrast with barcodes and readers, RFID is unobtrusive, scanning can be conducted at a distance, and the technology combines gracefully with wireless delivery of information from the Internet. Point a wireless device at an object with an EPC, and you can instantly learn all about it. RFID scanners might also be embedded in warehouse and store shelves for stock control, supermarket and library lanes for automated checkout, and your refrigerator for automated reordering of commodities.

Ubiquitous transponders and RFID tags have the potential to efficiently replace other forms of identifying and descriptive information that people and products carry—ID cards, labels, keys, passwords, and the like. Instead of reading a wine label, you might ping the label's embedded RFID tag to download detailed information from the maker's Web site to your wireless PDA and apply collaborative filtering to provide a personalized rating and recommendation. Instead of sticking something into a door to open it, you might just walk up, let the door interrogate your implanted RFID tag and check an authorization database, and have it automatically open.

But it is also vividly obvious that transponders and tags, combined with location tracking, could open the way to unwanted surveillance—of unprecedented precision, detail, and thoroughness. Since it is unlikely that they will just go away, they create an urgent need for sophisticated identity management systems that allow you to control exactly what you want to reveal about yourself, when you reveal it, and where you do so.[16]

URBAN INFORMATION OVERLAYS

The most profound effect of electronic tracking, inscription, and interrogation techniques is, in combination and on a large scale, to change the fundamental mechanics of reference—the ways in which we establish meaning, construct knowledge, and make sense of our surroundings by associating items of information with one another and with physical objects.[17]

The simplest way to associate information with physical objects is, of course, to inscribe text or images directly, and more-or-less permanently, onto their actual surfaces. This is the straightforward strategy of a manufacturer who puts a label on a soup can, a conference organizer who provides nametags, or a monumental mason who erects an inscribed headstone over a grave. It is also that of an architect who provides a directory at the building entrance, direction signs at circulation system intersections, and names on office doors; a visitor sees just what she needs, just where she needs it. For the philosopher— worried about the subtleties of sense and reference—these are among the elementary classroom cases.

Epigraphic technologies have evolved over time. Carved, painted, and ceramic letterforms and decoration have a long history. The electric light opened up the possibility of luminous inscription on a large scale—eventually producing the splendid excesses of the Las Vegas Strip in the 1950s and 1960s, as memorably celebrated in *Learning from Las Vegas*.[18] Since then, programmable electronic signs and decoration have gradually supplanted their simpler electrical ancestors. Times Square and the Ginza have become playgrounds of programmable LEDs, and Robert Venturi has suggested that the pixel is the new tessera. Looking back at the Strip in 1994, Venturi and

Denise Scott Brown noted: "The flamboyant neon of the Golden Nugget Hotel on Fremont Street has been removed, and on the Strip neon is being replaced by LED or its incandescent equivalents. The moving pixels permit changing imagery and graphics for a multicultural ethos in an information age."[19]

But direct inscription—durable or ephemeral, mainstream or marginal, conservative or subversive—is not the only option. You can also associate information with sites and objects by constructing a map, compiling a street directory, or writing a guidebook, and distributing copies to those who may want them. This works particularly well for making cities instantly legible to visitors; as long as they have a few readily recognizable landmarks to orient themselves, they can forget about trying to decode the local sign system and rely upon Lonely Planet, Access, or Zagat. Or if they want to go digital, they can try PDA-based city guides such as those provided by Vindigo.

The great advantage of these indirect information overlays is that they are nonexclusionary, there may be indefinitely many of them, and individuals can use whichever ones they want. They are always selective, incomplete, biased, and subjective, but there are always alternatives.[20] If Western visitors have maps and guidebooks in their native languages, it does not matter that Tokyo street signs and subway directions are in incomprehensible Kanji. If backpackers have guides to inexpensive accommodations and eating places, and affluent travelers have guides to international hotels and Michelin-star restaurants, members of each group can navigate as they wish. If you are straight, you need one guide to Provincetown, if you are gay, another; and if you are adventurous or confused, or just don't like the construction of these categories, you may want to map it both ways.[21]

Orientation is crucial, of course. Unless you can pinpoint your current position on a map, and point the map in the right direction, you remain lost. (Or, to put it more philosophically, it is not clear how the detached text in your hand *refers* to features of the surrounding environment.)[22] If you cannot connect that ambiguous ruin in front of you to the correct entry in your *Blue Guide*, the information that you have in your possession remains frustratingly useless. Even worse, you will retrieve inaccurate information if you don't get the spatial registration right.

You can keep track of your location on a map through some variant on the strategies of sighting recognizable landmarks (such as topographic features, landmark buildings, or stars), watching the time, and keeping careful records of your speed and direction—the elements of the ancient art of navigation. Traditionally, navigators have used compasses, sextants, and chronometers to assist with these tasks. Today, as we have seen, portable electronics are taking over. You can locate yourself on a map or at a relevant guidebook entry by means of GPS or other electronic location techniques.

GEOCODED BITS

Electronically generated coordinates are obviously useful if you have a map showing a coordinate grid; you can manually pinpoint your current location on the map. But they are even more so if you have access to a geographic information system (GIS)—that is, a database in which the items of information are associated with geographic coordinates.[23] You can then use search software to retrieve the items corresponding to your location.

One very common form of geographic information system is, in effect, an electronic street map. In the database, street corners and intersections are specified by their coordinates, and streets are described as links between such nodes. Graphic software converts this information into on-screen maps, and on request, software can compute shortest paths between specified street addresses. This is the basis of online map and direction systems, such as Mapquest. Combine this sort of GIS with a GPS receiver and you get an automobile navigation system that can continuously show your current location on a map displayed on a dashboard screen, compute the shortest path to your destination, and give you directions.[24]

Within buildings, facilities management (FM) databases provide digital floor plans, or even three-dimensional geometric models, instead of urban-scale maps. Typically, FM databases provide information not only about spatial configuration, but also about ownership, use, and condition of interior spaces, associated furniture and equipment, finishes, and services.

But street networks and floor plans are by no means the only kind of geocoded content. There is, in fact, a significant industry devoted to creating and maintaining specialized GIS and FM databases for a wide variety of purposes. Urban planners and marketers use GIS systems to map demographic and economic variables, geologists employ them to analyze mineral resources, utilities engineers use them to keep track of pipe and wire locations, and so on. Even more crucially, though, the distinction between specialized GIS databases and Web data in general is fading away; spatial metadata can be usefully associated with just about any Web content, and increasingly this is being done.[25] This can even be accomplished automatically, as when a GPS-equipped camera associates coordinates with every exposure, so that exposure locations can be plotted immediately on a map, and so that images can be retrieved and sorted spatially. In sum, spatially indexed databases can provide any sort of personalized guidance you may want, through any sort of terrain.

And since GIS and FM systems are digital, they can take advantage of capacious storage and efficient retrieval to provide instant access to vast amounts of information. This information may be cached in a vehicle or handheld device if it does not need to be updated too frequently. Or, if it is more dynamic, it may be pumped out continuously from a centrally updated wireless server. (Compare this huge, invisible store of instantly available information to the folded letter concealed in Leopold Bloom's hatband!) You can sort and filter this information according to your current interests—maybe, for example, using collaborative filtering to recommend and locate restaurants or bookstores on the street you happen to be strolling.

Furthermore, on-screen maps and plans are not the only sort of output. Automobile guidance systems, for example, can be set to provide sequences of spoken instructions—a particularly useful format when you have to keep your eyes on the road. Guidebook text can be geocoded for spoken output at the appropriate moment; this makes the text perform like an automated tour guide or docent—continually responding, in appropriate and interesting ways, to the current context—rather than a passive repository for looking things up.[26] Cities and buildings, like films, might be scored by famous composers—with the sound track electronically edited, on the fly, as you moved around. And *everything* of relevance to a particular location (for example, a historic site or a crime scene) might be retrieved and arrayed to provide a comprehensive, electronic mise-en-scène.

Different types of vehicles have different uses for geocoded bits. Shopping carts may trace customers' steps among the aisles, refer to their purchasing histories, and flash personalized purchase suggestions as they pass the corresponding shelves.[27] If you like cornflakes, and you haven't bought any in a while, you get reminded of that as you roll through the cereal section. Bicycles may respond to your current exercise requirements, routing you up hills or away from them, depending upon your instructions.

GIS and FM systems, spatial metadata, and associated technologies have mostly escaped the attention of social theorists and cultural critics.[28] They sound boring, and the large literature associated with them is mostly nuts-and-bolts technical. But don't be fooled; they represent the beginnings of an important new relationship between information and inhabited space—one that will become increasingly prominent as the wireless world evolves. Bits will increasingly be tagged with coordinates, and the digital doppelgangers of cities will be closely registered to their physical counterparts.

RETHINKING ACCESS

This emerging cross-linkage of the digital and physical domains allows us to rethink one of the traditional functions of a city—that of minimizing the time and effort its inhabitants must spend in locating and accessing resources and one another. Usually cities accomplish this through a combination of physical design strategies—high density, efficient transportation, and legible layout that reduces obscurity and confusion—with signposting, maps, and directories. Because urban configurations are relatively stable, inhabitants can, over time, build up mental maps that efficiently guide them to what they want. Information storage and retrieval systems, such as relational database systems or the World Wide Web, similarly seek to minimize search time, and make use of indexing, pointers, and efficient sorting and searching procedures to do so. The terrains and the means of traversal are different, but in each case the goals are much the same.

Now, as a result of cross-linkage, strategies of urban designers and database hackers are converging. Not so long ago, the inhabitants of cities physically searched for physical things, and electronically surfed for online information. Today, in addition, they may employ physical exploration and location-sensitive devices to get to geocoded digital information and take advantage of electronic guidance to assist in efficient location of physical things—from stolen automobiles to tourist attractions.

Furthermore, an appropriately programmed location-sensitive device can usefully reason about spatial and temporal information. If it knows your current location and your goal, it can compute the shortest path and cheapest path for you. If it knows the public transportation system schedules and fares, it can figure out the best way for you to go. If it knows your interests and time constraints, it can personalize tours for you. If it knows whom you want to meet, it can find and arrange a mutually convenient meeting place and, if necessary, help you to recognize each other. If it knows your schedule, your location, and the time, it can keep constant track of whether you are where you are supposed to be and discreetly remind you when it's time to leave for the next place.

SPATIAL REGISTRATION

GPS-powered, GIS-driven automobile navigation systems maintain a consistent relationship between the dashboard map and the surrounding terrain by keeping the map pointed in the right direction and keeping the vehicle's current location roughly in the map's center. The video displays provided to airline passengers, keeping them informed about progress toward their destinations, employ a similar strategy. This is a simple but effective way of registering computed information with the physical world—that is, of unambiguously associating it with the physical objects to which it refers.

A more sophisticated technique is to provide a heads-up display on the windshield. This superimposes graphic and textual information on the scene ahead, like virtual annotation or signage. And of course you no longer have to look down at the dashboard. So far, this sort of display has mostly been used in military aircraft (where looking down can be a particularly serious problem), but it is likely to become increasingly common in other types of vehicles.

The handheld or wearable equivalent of this is the see-through display, incorporated in a high-tech monocle, spectacles, field glasses, or a camera viewfinder. One type of see-through display is built around a miniature flat panel of light valves or light-emitting elements. Another makes use of a retinal scanning display; a pulsed beam from a low-powered laser diode is bounced off a tiny moving mirror, directed into the eye, and thus rapidly scanned across the retina.[29] Point an appropriately programmed see-through display at an object of interest, and it will tell you all about it. Maybe, as well, it will automatically add the commentary to a photo or video.

The very best spatial registration can be produced (in principle, at least) by combining accurate head tracking with stereo see-through goggles and a three-dimensional geometric model of the terrain. The result is a so-called "augmented-reality" system.[30] The technical challenges of accurate head tracking and spatial registration, and of constructing comfortable and unobtrusive see-through goggles, are formidable. But augmented-reality systems elegantly combine the advantages of direct and indirect inscription and have many potential uses. They might, for example, be incorporated into "smart hard-hats," and provide construction workers with x-ray vision of hidden utilities and structural elements. And construction drawings could be replaced by full-scale, correctly registered, three-dimensional "models in the air" to guide layout and on-site assembly of construction components.

Different contexts and purposes demand spatial registration of information in different ways and with different levels of precision. If you want to retrieve information about local weather or traffic conditions, accuracy to within a few kilometers is all you need. For the motorist, a correctly oriented and updated dashboard map, with accuracy to a few meters, may suffice for street navigation. For a construction worker, who must keep hands free and perform assembly operations with millimeter precision, a very high quality, head-mounted augmented-reality system may be necessary.

In any case, spatial registration combined with geocoding takes texts, images, and sounds from free-floating locations on pages and in storage devices and binds them to specific spatial and temporal settings. It dynamically overlays labels, commentaries, instructions, and abstractions on the physical objects, people, and places to which they refer, supplying precisely what the World Wide Web does not—context. Instead of the precession of simulacra that Jean Baudrillard promoted onto the cultural studies agenda in the 1980s,[31] you get site-specific installations of bits—hybrid constructions in which digital information adds a layer of meaning to a physical setting, and the physical setting helps to establish the meaning of the digital information. This adjoins a new dimension to architecture, and creates new opportunities for assertion of facts, construction of fictions, and insinuation of falsehoods.

ELECTRONIC MEMORY PALACES

Cicero, no doubt, would have loved all this. In *De oratore* he observed that orderly arrangement of information is essential to its memorization and that architectural space could effectively provide this arrangement. He illustrated the point with the tale of the poet Simonides, who remembered exactly where guests had been sitting at a banquet when the roof of the hall collapsed, mangling them all beyond recognition. Cicero went on:

> He inferred that persons desiring to train this faculty (of memory) must select places and form mental images of the things they wish to remember and store those images in the places, so that the order of the places will preserve the order of the things, and the images of the things will denote the things themselves, and we shall employ the places and images respectively as a wax writing-tablet and the letters written on it.[32]

This was the mythic origin of mnemotechnics—the art of memory—by means of which ancient orators could deliver long speeches from memory. Frances Yates has summarized the procedure, as described in detail by Quintilian:

> In order to form a series of places in memory, he says, a building is to be remembered, as spacious and varied a one as possible, the

forecourt, the living room, the bedrooms, and parlors, not omitting statues and other ornaments with which the rooms are decorated. The images by which the speech is to be remembered—as an example of these Quintilian says one may use an anchor or a weapon—are then placed in imagination on the places which have been memorized in the building. This done, as soon as the memory of the facts requires to be revived, all these places are visited in turn and the various deposits demanded of their custodians. We have to think of the ancient orator as moving in imagination through his memory building whilst he is making his speech, drawing from the memorized places the images he has placed on them.[33]

In today's emerging electronic mnemotechnics, information is stored in digital devices rather than heads, it is associated with physical places through geocoding, it may be retrieved by *actually* moving from place to place, and it may be presented in multimedia format on devices such as see-through video displays and audio earpieces.[34] In this fashion, a whole city becomes a vast, collectively constructed memory palace that divulges its contents to inhabitants as they circulate through it.

BITS COME HOME

Long ago (somewhere around 1995), as today's digitally networked world was big-banging into existence, it sometimes seemed that cyberspace might imitate, improve upon, and ultimately supplant physical space as the place to be. Point-and-click interfaces to personal computers were luminous replicas of physical desktops. Virtual communities represented themselves on-screen as cozy little villages—nicer, less problematic places than the neighborhoods most of us bodily inhabited. Among designers, CAD systems and digital models were rapidly replacing physical media. Computer graphics researchers, in a SIGGRAPH-by-SIGGRAPH progression that recalled the development of Western painting from Masaccio to Manet, sought ever-more photorealistic renditions of physical scenes—maybe moviemakers would no longer need real actors and sets. Videogames and virtual reality environments were populated with humanoid avatars and three-dimensional models of familiar physical objects. Cyberpunk novelists imagined shucking your messy flesh for a perfected virtual body and dwelling among the pixels. The Internet, we were assured, changed everything—distance was dead, the economy was weightless. Maybe (in a new twist on ancient yearning to transcend corporeality) we would live forever in reliably engineered silicon.

But at a more structural level, a complementary transformation was stealthily taking hold; physical space was acquiring many of the crucial characteristics of cyberspace. Material things, with embedded computation and communication capabilities, were starting to function like on-screen graphic objects; poke them or move them, and you got some computational result. Through location tracking, physical artifacts could report where they were, much like cursors in screen coordinate systems. Through wireless communication, bodies, places, and devices could be as densely, continuously, and seamlessly interconnected as Web pages. Online information—no matter where it physically resided—could show up wherever and whenever you needed it. By the time the dot-com bubble burst and splattered, it was clear that physical space and cyberspace had actually become locked in an intricate, mutually transforming embrace, with functions shifting and dividing between the two in complex ways.

By the early 2000s, bits had returned from cyberspace. They had gone on location in the material world.

8

FOOTLOOSE FABRICATION

We produce the artifacts we want by bringing together designs, energy, and materials—all of which may be delivered to the production site, from a distance, through networks. When you bake a cake in your kitchen, for example, you follow a recipe that may have arrived through the mail, you apply thermal energy that was most likely supplied in the form of piped gas, and you combine ingredients transported from around the world through a variety of transportation networks. When you laser-print a piece of artwork in your office, the design arrives over a computer network, the electricity to activate the print mechanism is supplied by the grid, and the paper and toner cartridges (the weak link in this web) probably arrived by truck.

Very few artifacts are produced directly, at one location, from natural raw materials. Most are in fact put together from other artifacts; in other words, they emerge, through multistep processes, from supply networks. The nodes of these networks are sites at which materials are organized, according to some design, through the controlled application of energy. Today—notably, for example, in the fabrication and assembly of electronic components—these networks may extend globally, and production usually requires careful timing and coordination of parallel activities at multiple sites.

In the era of craft production, multiple fabrication and assembly tasks were performed sequentially, at one location, in the craftsman's shop—a strategy celebrated, for example, by William Morris. Industrialization sought the advantages of division of labor, specialization, and parallel processes; industrial assembly lines form transportation networks linking sites of specialized tasks. In today's network era, enhanced transportation and telecommunication capabilities allow greater spatial division of labor and, in effect, the expansion of assembly lines from factory to global scale. If you are a twenty-first-century producer, you don't just manage plants, you manage complex supply networks.

DECENTRALIZED PRODUCTION

When the machines that bring designs, energy, and materials together at nodes in supply networks are bulky, heavy, and expensive (as with high-speed printing presses and CD burners), there are few of them, and their locations tend to be fixed. They are potential bottlenecks in the flow of production. Furthermore, as robber-baron capitalists and Marxist revolutionaries knew equally well, they provide opportunities to acquire political power by grabbing the means of production. But when miniaturization allows the development and proliferation of small, inexpensive production devices (such as laser printers), redundancy can be introduced into supply networks. Production can often be decentralized and even mobilized. Under these conditions, political power can be distributed (and centralized power can be subverted) by producing such devices in quantity and spreading them around.

Consider, for example, the evolution of the supply networks through which the humble ice cube cometh. The frozen water trade began, on a large scale, with the harvesting of big blocks of ice from New England lakes and rivers by means of ice plows.[1] Increasingly efficient long-distance bulk transportation networks made this possible. Blocks were stored in huge icehouses near the rural harvest points, transported by sailing ship to cities as distant as Calcutta, stored again in urban icehouses, and eventually distributed by ice cart to domestic iceboxes. By the 1880s large, steam-driven artificial ice plants were becoming competitive—particularly in warm locations far from the sources of natural ice; these depended upon water supply, energy supply, and local delivery networks, and they were a first step in decentralization of ice production. Then came electric supply networks, small electric motors, the invention of sealed-container refrigeration, and mass-produced domestic refrigerators. Ice production had been a centralized industrial activity, but by the 1950s it had fragmented and recombined with domestic space. Today small, automated ice-makers are nodes in domestic electrical and plumbing networks, and bite-sized ice cubes pop out of your refrigerator door; the distance from production point to your glass has shortened from thousands of kilometers to a few centimeters.

Limits on network capacity, costs of transporting things through networks, and the losses incurred in transportation also play roles in establishing the relative advantages of centralized and decentralized production. Steelworks are often concentrated near sources of iron ore and coal, for example, since it is more expensive to transport these commodities over long distances than to transport the less bulky finished products. Ice plants were located near ice markets, since ice gradually melted during transportation. And the mills of the early industrial era were clustered around sites of water or steam power, since production machines had to be within range of the belts and other mechanical means that were used for transmitting power. But when the crucial networks become ubiquitous and efficient, as with modern electrical grids and the Internet, the importance of distance correspondingly diminishes; it hardly matters where you plug in a PC to download and print a document from a Web server.

These effects are most pronounced, and perhaps most threatening to established industries, when mobilized, easily replicable software transforms large numbers of Internet nodes into a highly redundant, geographically dispersed, production and distribution system. Thus when music industry lawyers tried to shut down the KaZaA peer-to-peer file-sharing system, they found that the developers were living in the Netherlands, the programmers were in Estonia, the location of the source code was unknown, the distributor was based in Australia but incorporated in Vanuatu, and installations were scattered over 60 million Internet users in 150 countries.[2] In the long run, the established music industry had about as much chance as the New England ice harvesters.

PERSONALIZED PRODUCTION

Traditionally, industrial production has sought scale economies. To compete effectively, industrialists built big, fast machines that turned out long runs of standard products at the lowest possible cost. (The higher the investments in this machinery, the more intensively it had to be used.) But decentralized, personal production is more concerned with personalization—that is, with smaller runs of products that may be more expensive but are more precisely adapted to the specific requirements of particular contexts. It is the difference between a high-speed newspaper press and your personal laser printer.

You can provide for customization by equipping a production device with lots of knobs and dials. A really good manual camera, for example, is festooned with gadgets you can adjust and set, and provides a photographer with extraordinary control of the subtle qualities of images. A point-and-shoot camera, by contrast, efficiently produces a highly standardized product and, since it doesn't have to provide so many external control points, is usually a smaller, cleaner, and simpler-looking object.

As production devices have acquired embedded intelligence, a new possibility has emerged; you can control (or hack) them by sending them streams of bits. So your laser printer can produce different pages, and your MP3 player can produce different sounds, depending upon the digital input. At another level, if you have the programming skills, you can mess with the resident code that interprets such input—so altering the types of products that are output.

At more industrial scales, numerically controlled (NC) production machinery has gradually been taking over from older, more cumbersome, manually controlled devices. At first, NC machinery was driven by paper tape; now it is run directly by computers. The shift to NC reduces setup costs and the need for close manual control and so allows production of varied output without crippling cost penalties. Thus a numerically controlled laser cutter can efficiently produce highly varied shapes from sheet material, and a numerically controlled deposition printer can produce three-dimensional solids. By the 1990s this shift was vividly being reflected in architecture; buildings like Frank Gehry's Guggenheim Museum in Bilbao were no longer constructed from the simple, repeating components that had characterized construction in the industrial era, but from complex, CAD/CAM-fabricated, nonrepeating elements.[3]

As inexpensive, digitally controlled production devices have been networked, it has become possible to download designs from distant servers, maybe tweak and customize them as required, then produce them locally. So increasingly it makes sense to distribute designs through telecommunication networks rather than finished products through transportation networks and to produce customized physical artifacts on demand, wherever and whenever they might be needed.

Fax machines provided an early, modest intimation of this new logic of decentralized, customized, production at a distance. At the transmission end they begin by harvesting text and images—the graphic design that is to be reproduced at a distance—from paper sheets. They encode and send this information electronically, at high speed. At the reception end, they reinscribe it on new paper sheets.

When you fax a document you don't *actually* perform the miracle of teleporting a sheet of printed paper, but the effect is the same. A good Platonist would know just how to describe the process. You separate an object's form from its material, transmit the dematerialized form, and eventually reembody the form with new but indistinguishable material. You dematerialize, then rematerialize. You keep the bits constant, but substitute new atoms.

Even more dramatically, since laser printers have become commonplace, vast quantities of textual and graphic material are now stored on servers, downloaded anywhere there is a network connection, and printed out on demand. The distribution system for printed documents has, in effect, bifurcated; instead of printing documents at a central location then distributing information and paper bonded together, the idea is to distribute blank paper and ink as inexpensive commodities and to distribute information separately via highly efficient electronic channels. In many contexts, particularly where the numbers of copies are small or the information is frequently updated, the efficiencies of this distribution strategy far outweigh the economies of scale realized through centralized high-speed printing.

This has, of course, crucial implications for publishers, book retailers, and libraries. Ever since the industrial revolution gave us high-speed presses, books have been produced in bulk at central factory locations, stored (at high cost) in warehouses, distributed to retailers who store them again on their shelves, and finally delivered to libraries and consumers. This system makes it expensive and difficult to keep titles in print for long periods and to distribute books effectively to remote corners of the world. Furthermore, lots of books are distributed to retailers only to remain unsold, eventually to be returned to the publishers and pulped. Now, however, it is feasible to store books in electronic form on servers rather than in print form in warehouses, to download them on demand to point-of-sale book machines, and to print and bind them on the spot.[4]

And the potential of new print technology does not end there. At MIT's Media Laboratory, Joe Jacobson and his research group have prototyped the idea of a "desktop fab."[5] The idea is to print logic chips, electronic tags, MEMS devices, and the like on inexpensive substrates such as paper or plastic using "ink" consisting of nanometer-sized particles. Instead of purchasing standard electronic components that were produced in billion-dollar specialized facilities, you might download and print your own—much as you now download software.

A NEW LOGIC OF PRODUCTION

As a result of decentralized, customized, production at a distance, supply networks are destabilized and transformed, and they demand rethinking. From an economist's perspective, for example, a chain through a supply network progressively adds value to materials—going, for example, from water to ice cube, sand to a silicon chip, or electronic components to fully assembled computer. Traditionally, the sequence has been one of design, component fabrication, assembly, warehousing, retailing, and eventually disposal or recycling. Now the producer's crucial assets may be designs residing on servers rather than completed products in warehouses, and these designs may be applied to add value to materials at multiple, decentralized production points of fluid and often indeterminate location, rather than at fixed, centralized industrial sites.

From a designer's perspective, the process is one of progressively binding design decisions to materials. At each stage, designers are constrained by the ways in which variables were bound at earlier stages. Thus if I assemble a design out of Lego blocks, I am constrained by the predefined geometry of my components; if I assemble an electronic system out of standard chips, I am constrained by the earlier decisions of circuit designers; and if I assemble an order for personalized sneakers online, I work within a tightly bounded range of sneaker features and their combinations.[6] But we are now moving from early binding of design variables to late binding. As a designer working with a networked, digitally controlled production device, I can select the design to download and customize it to my circumstances by choosing values for parameters at the very last moment before impressing it upon materials.

It matters little whether the digital master files that increasingly control decentralized production of material and acoustic artifacts are generated by processes of scanning, recording, and otherwise capturing existing physical reality, whether they are constructed directly by means of text, music, or CAD editing software, or whether they are complex hybrids. In a digitally controlled world, Nelson Goodman's hitherto telling distinction between autographic and allographic production collapses. Every artifact has a digital script, score, or plan, and every production process is an automated performance. Creation of the script, score, or plan may look like composition (taking place in meticulous, step-by-step fashion, using some sort of editing software) or more like recording a performance—with the operations executed at high speed, using specialized instruments and interfaces.

And this produces a curious new category of art objects. Under conditions of craft production, art objects are unique. We value their direct connection to the hand of the artist, and we care about provenance; they possess the quality that Walter Benjamin famously termed "aura."[7] Under conditions of mechanical reproduction, there is a distinction between an original (an oil painting by Vermeer, for example) and its copies, such as the postcards for sale in the museum store—a mass audience consumes these aura-challenged, industrially produced commodities. But under conditions of materialization on demand from a digital file, there need be no original; new material instances may be produced at any time, and instances may differ widely from one another—for example, by being produced at different scales, at different resolutions, or with different materials.

The shift to decentralized, digitally mastered production is inexorably eroding structures of authority that had been sustained by mutually supportive strategies of productizing intellectual and artistic work, industrializing mass production, and legally controlling replication processes. The debate buzzing around this transformation has mostly been cast in terms of ownership of intellectual property, payment, copying, and reuse rights.[8] It has pitted the aggressive lawyers of large entertainment and publishing companies against libraries, academics, and defenders of the public interest in the areas of fair use and unrestricted access. But maybe even more fundamentally, the debate is also about preserving the stability, identity, and closure of intellectual products versus the possibility of creative transformation and recombination.

If you purchase a CD from a record store, for example, you get the set of performances that the record producer has chosen to bundle together—no more, no less—whether you want them all or not. But if you rip all your CDs, and store the MP3s on your hard drive, you can group and sequence performances in any way you want.[9] And if you download MP3s from Napster or one of its successors, you gain even more freedom.

This breaks down the unity of the CD while preserving the integrity of recorded performances, but the musical strategy of sampling takes the process a step further. The sampler employs digital editing technology to appropriate fragments of musical material from multiple sources, transform them, and recombine them to produce new works. This is an innovative and artistically vital practice but, under the usual rules of industrial-era copyright, it is treading on dangerous legal ground. Artists get away with it only so long as the appropriated fragments are not copyrighted by aggressive holders, not overly large, and not too recognizable after transformation. If their compositions turn out to be profitable enough to make it worthwhile, they are likely to find themselves pursued by IP sharks.

Print publishers, like record companies bundling tracks on CDs, have traditionally grouped texts into issues of magazines and journals and into bound hardback and paperback volumes. In addition to its efficiencies and pricing advantages, this provided a convenient framework for branding; it was advantageous to have your work come out under the imprint of a well-regarded publisher or to have your article published in a prestigious journal in the company of the established intellectual elite. The photocopier first challenged this strategy by enabling ready reproduction of pages, articles, and chapters, and the recombination of these extracts into new, ad hoc collections such as course readers. With digital text, the logic of the database replaced that of the printed page. Publishers of online journals discovered that rigid subdivision of material into "issues" no longer made much sense. Large, searchable text databases like LexisNexis aggregated articles from numerous sources and supported retrieval by user-specified category, and the World Wide Web cross-connected huge structures of text with hyperlinks.

For a century and a half photographers, using the silver-based process, have captured complete images with definite spatial and temporal coordinates. The privileged role of the photographic image as reliable visual evidence has depended upon its wholeness and closure and the possibility of tracing it back to an unambiguous origin point. But digital cameras now decompose images into finite grids of pixels that may readily be sampled and recombined to produce seamless collages. Furthermore, digital images—both captured and synthetic—may be replicated indefinitely and endlessly circulated through the Internet. Distinctions among visual facts, falsehoods, and fictions are increasingly difficult to construct and sustain.[10]

From the perspective of architects, Napsterization is the culmination of a long process of mobilizing and recombining design information. It began with the use of portable templates to facilitate replication of standard shapes and profiles in buildings. With the emergence of print, architects began to publish descriptions of architectural elements and rules for their combination; great classical treatises, from Palladio's *Four Books* to Guadet's *Elements and Theory*, disseminated standardized languages of architectural form.[11] We are now entering an era in which descriptions of elements and rules are stored on servers as software objects, traded, varied, and recombined electronically, and eventually materialized by means of CAD/CAM production devices.[12] If Palladio were alive today, he would be looking to 3D digital modeling and peer-to-peer distribution technology, not to woodcuts of plans and elevations.

MODULARITY AND PORTABILITY

You can trade and recombine MP3 files, digital images, CAD models, and the like because they are modular; that is, they are standard units in consistent format. (If an MP3 player gets a file that isn't in this format, it won't open it.) More technically, the differences among MP3 files are hidden behind a layer of abstraction that allows them to be handled in a uniform way. Over time, computing environments have evolved increasingly sophisticated layers of abstraction, providing greater modularity, easier reuse, and enhanced capacity to recombine digital fragments into new structures.

In the 1960s, on the batch processing computers of the time, you could assemble Fortran programs by duplicating decks of cards encoding functions and subroutines, then shuffling them into desired sequences; rubber bands, sorting machines, duplicating machines, and cardboard boxes were important aids in this process. Similarly, data files were decks of cards terminated by punched end-of-file symbols. If you wanted to modify a line of code or a data record (to correct a typo, for example) you physically pulled a card and replaced it. But this was cumbersome, and furthermore, mutually incompatible compilers established strong "trade barriers" among computing subcultures centered on different machines.

By the 1970s, though, things had become much easier; at the terminals of timesharing mainframe systems, you could use simple filesharing systems, online function and subroutine libraries, and text editors to assemble and run programs (maybe in Lisp) electronically. Then local area networks, the Arpanet, and the Internet further facilitated sharing, reuse, and recombination of code—and provided fertile ground for development of the collaborative "hacker" culture. Meanwhile, programming languages and software engineering practices evolved to support creation of modular, portable, reusable code units rather than huge, monolithic software constructions of the past; in particular, languages like C++ enabled creation of highly modular software "objects," development of sharable object libraries, and use of convenient mechanisms, such as inheritance, to facilitate modification and combination of existing objects to produce new objects.

Eventually, as networked computing environments became standard, languages like Java combined the virtues of modularity with easy network distribution and ability to run immediately in just about any computing environment. At a technical level, there was now little to inhibit global free trade in software modules. Of course, though, you might still choose to inhibit such trade by deliberately introducing incompatibilities, or to erect security barriers against ill-behaved code you wouldn't want invading your machine.

Modularized, parameterized, mobilized software has enabled several competing production and distribution strategies. For example, the neo-Fordist, industrial strategy—as pursued by Microsoft and other large software firms—emphasizes organized division of labor, accumulation of corporate intellectual property, integration of as many functions as possible into a single, standard, product, closure and protection of that product (the source code is not available to users), branding, and market domination.

By contrast, the open source strategy—most vividly illustrated by the development of the Linux operating system—takes advantage of the creativity and enlightened self-interest of user communities to create shared intellectual capital.[13] In an open source production environment, source code is made freely available, users extend and modify it as appropriate to their particular needs and priorities, and contribute generally useful extensions and modifications to a common pool.

Most radical of all (and, so far, of least practical impact—though I would not bet against it in the long term) is the evolutionary strategy. In so-called "simulated evolution," software modules randomly mutate, they are evaluated by some specified fitness function, and according to their fitness values, they either survive in the common pool or are discarded.[14]

All of these strategies can work well under the right circumstances, and they can work in various combinations and flavors; the commonality is reliance on modular, malleable, mobile, electronic text as the enabling medium.

These network-enabled strategies of decentralized collaging, sampling, searching and exchanging, and open sourcing threaten established corporate approaches to creating, pricing, marketing, and protecting information products. Not surprisingly then, corporate interests have often resisted and even attempted to criminalize them.[15] Disney, Time-Warner, Microsoft, and Reed-Elsevier would have it that the value of digital information derives from easy distribution and wide consumer access to completed, packaged, impregnably encrypted and copyrighted intellectual products. But far greater societal value resides in its endless capacity for fluid adaptation, transformation, and recombination, within communities of interest, to produce new and unexpected outcomes. If publishers and record companies succeed in enforcing continuation of outdated, industrial-age norms and conventions into the network era, many of the advantages of free-flowing, readily recombinable digital information will be lost. The cultural cost will be enormous.

This applies not only to pure information products, but also to material products formed through the application of information. The binding of form to materials has been loosened. We are entering an era in which batches of material may readily inhabit different forms, and digitally specified forms may inhabit different materials.

In sum, we have to rethink Manufacturing 101 and reconsider strategies for controlling the means of production. We may still be a long way from the superhero nano ring or the future wittily extrapolated in Neal Stephenson's *The Diamond Age*—a world in which every household has "the feed," a nanopipeline that supplies atoms to matter compilers that produce whatever goods you need on demand—but conditions have fundamentally changed.[16] What now matters most is not having an inventory of valuable *things* in your possession, nor even the machinery needed to produce such an inventory, but *access* to the invisible, immaterial, digital specifications. It is all very Platonic, in a way; digitally encoded ideas exist somewhere in cyberspace, and physical artifacts are their imperfect, material realizations.

9

POST—SEDENTARY SPACE

Antonello da Messina's famous picture of Saint Jerome in his study shows how the attachment of atoms, in the most literal way, adds inertia to information.[1] It depicts the scholar surrounded and encumbered by his heavyweight personal information environment; he pores over a manuscript on the inclined desk before him, and his precious accumulation of books, papers, and writing instruments is gathered around, within the enclosure of his room. It is a tightly bounded assemblage of physical materials at a particular location. If he wants to keep it at hand, rather than rely upon fallible memory, its bulk and mass tightly tether him to that particular place.

Dilbert wasn't so different. Like most office workers of the 1980s and 1990s, he occupied a cube containing a PC.[2] His information environment was more digital than paper—some of it residing on his local disk, and some of it on distant servers. Although he did not need physical proximity to the servers, he still had to be near the delivery point. He was as tied to his computer as Saint Jerome had been to his bookshelves. Furthermore, his telephone was physically attached to the cubicle; you could only call him when he was in. In other words, desktop computers, telephones, and wired networks provided fixed *points* of presence. These favored points were like oases in a digital information desert; they were powerful attractors of human presence and activity.

The cubicle farms of the era were grids of such points. As networks became faster and more sophisticated, and as more and more information moved from local storage to servers, it ceased to matter which cubicle Dilbert occupied. He could log in from wherever he happened to be in the building, and he could have his calls electronically redirected. This was good for the boss, who could make more efficient use of his stock of cubicle space by shuffling cubicle-dwellers around at will, but it did not help Dilbert's quality of life. He still had to be at *some* point of presence, and the new cube-to-cube mobility gave him even less opportunity to personalize his surroundings.

But wireless connections and portable access devices create continuous *fields* of presence that may extend throughout buildings, outdoors, and into public space as well as private. This has profound implications for the locations and spatial distributions of all human activities that depend, in some way, upon access to information.

FIELDS OF PRESENCE

By selectively loosening place-to-place contiguity requirements, wired networks produced fragmentation and recombination of familiar building types and urban patterns.[3] For example, the local bank or branch bank largely disappeared in the early digital era; it was replaced by more decentralized access points distributed throughout the city—that is, ATM machines and electronic home banking on desktop computers, combined with centralized back office facilities and call centers that provided economies of scale. Similarly, by selectively loosening *person*-to-place contiguity requirements, wireless networks and portable devices have created an additional degree of spatial indeterminacy; you can now electronically "home" bank from a wireless laptop, and if you can rely upon credit cards, debit cards, or some form of electronic cash, you never need to go looking for an ATM location. From the customer's perspective, banking no longer has any particular place in the city.

In the short-lived dot-com era, Amazon and other online retailers fragmented and recombined retail functions and provided an alternative to stores and malls; you would order goods online, and they would be delivered from a warehouse to your home or office. Thus the retail store functions of advertising and enticement, browsing, and completing the sale transaction were fragmented and dispersed, while the functions of storage and dispatch of stock were highly centralized at monster-box (often national) distribution centers that offered economies of scale, and back-office functions—supported by e-commerce technology—could be located just about anywhere. This turned out to work well for compact, high-value items like books and electronics, but for dog food, delivery costs were a killer. Now, wireless fields of presence are beginning to provide yet another alternative—one that returns advantage to highly differentiated, place-based retailers. Location-based advertising, maybe combined with electronic urban navigation, can tell you the nearest point of availability of some specialized thing that you want—such as a funghi porcini risotto at a good Italian restaurant—and guide you to it. If you think of shopping as a search for scarce or specialized (maybe unique) things that are to be found at particular locations in a city, then wirelessly delivered, dynamically updated, location-specific information can greatly increase the efficiency of that search.

Location-based advertising can also guide you to scarce commodities available at dynamically varying locations, such as urban parking spaces and unoccupied machines in laundry rooms. This can be integrated with sensor networks; you can embed inexpensive, car-sensing devices in parking stalls, and use monitors in washers and dryers. For example, eSuds has networked washers and dryers in college dorms; when machines are available, and when they have completed their cycles, they send out email and pager messages.

Some activities, such as stock trading and gambling on horse races, depend upon highly time-sensitive information. Before high-speed telecommunication, this meant that you had to be right at the site of production of that information—the stock exchange floor or the racetrack, for example. If you were distant, you might still participate, but you would be disadvantaged by information transmission delays; conversely, as James Joyce's get-rich-quick, wireless-aided scheme for betting on the English races illustrated, you might gain advantage by shortening those delays. With telecommunication, and electronic information and transaction services, the action began to shift to distributed points of presence; we saw the decline (and sometimes disappearance) of centralized exchanges and clusters of bookies on the rails, accompanied by the rise of electronic trading floors, day-trading at PCs, and off-track betting shops. Today, even those distributed sites are no longer technically necessary—though they may sometimes be preserved by regulators who want to confine activities to particular locations; trading and gambling can readily shift to wireless, portable devices.

Yet other activities have begun to depend upon clandestine, wirelessly distributed information and indeterminate locations. Where prostitution is legal (or at least tolerated), for example, brothels and red-light districts can be at fixed, well-known, maybe even advertised locations; where it is frowned upon, and telecommunication is not readily available, you get mobile streetwalkers; and where a wired telephone network is available, you get call girl operations—effectively mobilizing and distributing the actual sites of service but leaving the headquarters vulnerable to raids. But where clients, pimps, and sex workers have access to pagers, cellphones, and instant messaging, locations are temporary, mobile, and indeterminate. It becomes much harder to regulate the sex industry—and to protect workers.

Drug dealers have also learned the advantages of continuous fields of presence. Where they can rely sufficiently upon fortification, intimidation, and paying off the cops, they can distribute from crack houses. If a neighborhood has plenty of public telephones, they can arrange more dispersed transactions through those—but the authorities may stake out or simply remove telephones. When they have pagers and cellphones, though, dealers can stay highly mobile, take orders wherever they happen to be, and deliver at ad hoc locations—dropping rocks from an SUV and using foot action for ziplock baggies of weed. In response, during the 1990s many schools banned wireless electronic devices.[4]

Pornographers, as they always managed with new technology, have been particularly fast on their feet. Photography and print facilitated the production of pornography but generated a distribution problem; the obvious, site-specific option of a porn store was vulnerable to busts and shakedowns, might embarrass the customers, and might be resisted by neighbors. Pornographers quickly turned to the smut magazine format, the mail, and plain brown wrappers, but they still had to contend with mail regulators and customs officials. (Even today, don't try smuggling skin pix into Saudi.) They swiftly took advantage of the Internet, offshore servers, encryption, and anonymous remailers, but the authorities immediately countered by raiding server locations and looking for downloaded nasties on PC hard drives. Now it is possible to create highly redundant, decentralized, peer-to-peer porn networks (that is, to Napsterize smut), and to distribute to wireless devices at mobile locations. As broadband wireless develops, delivery of high-resolution pornographic images and videos (maybe captured live by wireless camera/transmitters) will undoubtedly be a killer app.

Skeptics about wireless networks like to emphasize that wireless merely provides the last link in a telecommunication infrastructure that is, in fact, mostly wired. Furthermore, it is a relatively unreliable and inefficient link. They are right, but they completely miss the point. As these examples illustrate, for good or ill, continuous fields of presence provided by wireless networks can fundamentally alter patterns of resource availability and space use.

REMOBILIZING SERVICES

The simplest of transportation networks allowed itinerant poets and scholars to visit communities and doctors to make house calls. However, as educational and medical services increasingly depended upon accumulations of specialized equipment, supplies, and expertise, they were centralized at large-scale, purpose-built facilities—particularly modern schools, university campuses, and hospitals. Sometimes students and patients became long-term inmates of these facilities, sometimes they visited them on regular schedules (school days and school hours), sometimes they went by appointment, and sometimes they made emergency visits—but always they had to remove themselves from the contexts of their communities and enter particularized environments to access the services they needed. The ivory tower and the magic mountain came to symbolize this system's reliance upon separation, the invigilated exam room and the Nurse Ratchet its engagement with structures of control.

Throughout the twentieth century, educational and medical complexes (sometimes combined) grew and coevolved with urban networks. Sometimes they were simply concentrations of specialized facilities within the urban fabric, as at University College London or New

York University, and sometimes they were discrete, clearly bounded, even walled and gated campuses. Increasingly extensive transportation networks enabled them to serve larger populations, provided their personnel with access to housing, and encouraged them to take advantage of scale to develop specializations. They evolved into major urban elements and important nodes in transportation networks. The UCLA campus, at the edge of the Los Angeles basin, may be the mature masterpiece of the genre; there is a pretty piazza at the center, the pavilions of the arts and humanities occupy the green and hilly northern end, big-footprint medical and engineering buildings cluster at the more urban southern end, and the perimeter is defined by a ring road studded with huge parking structures and card-key access points. Daily commuters converge on campus from throughout the basin, the San Fernando Valley to the north, and Orange County to the south.

The Arpanet—which, fittingly enough, first took root at UCLA—initiated a process of integrating these institutional nodes into even larger-scale networks. It was initially conceived (or at least justified) as a mechanism for sharing expensive computer resources among campuses and research centers; maps of early Arpanet growth show rapid proliferation of campus-to-campus links. Over time, computers got cheaper, campus computer centers (powerful political entities in the 1960s and 1970s) fragmented into campus LANs such as MIT's Athena network, and the Arpanet evolved into the Internet. Long before the dot-com boom of the late 1990s, the dot-edu domain was linking campus offices, laboratories, and dorm rooms globally. Furthermore, the policy on many campuses was to pay interconnection costs centrally rather than to charge individual users, so the Internet seemed like a resource that could be used freely.

At the macro scale, during the 1980s and 1990s, wired interconnections encouraged division of labor among campuses, increasingly ambitious and long-distance campus-to-campus collaborations, and formation of geographically distributed professional communities. Much as designers of the Arpanet had focused on long-distance sharing of computational resources, college, university, and medical center administrators became increasingly interested in the possibility of sharing specialized human resources through classrooms and lecture halls equipped for videoconferencing, telemedicine suites, and Web sites. At the micro scale, though, wired networks tied teachers and students to fixed points of presence—often to the detriment of face-to-face community; they tended to keep faculty members (like Dilbert) in their offices and students in wired dormitory rooms. As local and distributed communities competed for mindshare, local began to lose ground; professors and medical specialists often discovered that they had more valuable and satisfying exchanges with colleagues on the far side of the world than with those just down the hall.

Meanwhile, the Internet had expanded far beyond its ancestral heartland in academia—making not only on-campus sites, but also homes and offices generally, potential delivery points for remote education and telemedicine services. In their most elementary form, remote service sites simply pumped out information that might otherwise have been delivered through print, lectures, or face-to-face consultations—maybe personalized according to the stored needs profiles of clients. With higher bandwidth and more sophisticated software, these sites could efficiently provide interactions that approximated face-to-face; you might take a test online from the comfort of your home, or if you lived in a remote community, you might get a reasonably good medical examination through videoconferencing. And with embedded, networked sensors and intelligence, remote sites might perform highly specialized functions that, traditionally, had only been available at central facilities. Homes, for example, could become sophisticated medical monitoring and data collection nodes.

Finally, wireless networks superimposed continuous fields of service presence onto this pattern of major and minor points of presence. In particular, the introduction of campus wireless networks in the early 2000s, combined with portable wireless devices and the growth in electronic distribution of educational material, quickly began to break down the rigid person-to-place connections that had hitherto characterized campus life; students did not need to be at fixed network drops or computer clusters to download or electronically interact, they did not have to go to the library to pick up texts, they no longer required desks or carrels to write, and they did not need to show up in person for videocast lectures.[5] They could form ad hoc discussion and collaboration clusters, wherever and whenever they wanted, without losing access to online resources or contact with their geographically extended communities. The classically minded recalled that the Greek philosophers did not have offices and classrooms elaborately organized into departments and schools; they strolled freely through the groves and stoas of academe—but, as Plato vividly emphasized, they had to rely upon their memories. Now, while a scholar stands in line for a sandwich, she can wirelessly search the *TLG* for a passage from Plato to clinch an argument with her lunch companion.

As wireless education serves the mind, wireless medical care attends to the body—and the effects of the resulting service continuity are, potentially, even more dramatic. When monitoring and emergency summoning devices began to migrate from the hospital to the networked home, continuity of service could be increased, and separation of patients from their families and communities could be decreased. Now further migration, to wearable and implanted, wirelessly connected devices, is taking that process a step further. And more active care, through wirelessly operated implants and medication delivery devices, is an obvious possibility.[6] All this is not so different from continuously monitoring the conditions and dynamically responding to the biological needs of deep-sea divers or astronauts— but without the cumbersome suits and the tethers back to the ships.

Marketers of goods and services can potentially make use of wireless, context-aware devices to monitor customer needs continuously and respond to them appropriately, thus gaining a competitive advantage.[7] If you are in a theater, for example, your cellphone might discreetly vibrate in your pocket instead of ringing loudly and disturbing your neighbors. If you are on foot and it's raining, you might get prompted with the telephone number of the local taxi service and locations of nearby establishments selling umbrellas. Increasingly, businesses will try to be wirelessly present and responsive wherever, whenever, and however their customers need them—but this will, of course, depend upon the willingness of their customers to give up some privacy in order to gain personalized, contextualized service.

The advantages of wireless fields of presence are accompanied by subtle and not-so-subtle challenges to the regime of separation and control that has long been built into schools, campuses, and medical facilities. Schoolteachers were among the first to notice this, as kids equipped with wireless devices began to order pizzas in class, pass SMS notes, and clandestinely circulate answers to test questions. Retailers began to worry that customers in their stores might scan the barcodes on products and wirelessly search for competitors with better prices. Professors delivering lectures began to wonder if they had the undivided attention of students crouched over their wireless laptops and were surprised in seminars as their interlocutors silently downloaded salient information to interject into the discussion. The immediate authoritarian response, of course, was to try to banish personal wireless devices—but this was doomed to failure. When the dust settled, few would willingly give up the benefits of continuous connection.

RESTRUCTURING LIVE/WORK

Before the emergence of large-scale networks, dwellings and workplaces were often intermingled in fine-grained spatial patterns. Shepherds slept with their flocks, peasants lived among their fields, craft specialists both lived and worked in urban professional quarters, and merchants resided above their shops. But as Engels so vividly observed in Manchester, industrial cities changed that. Factory owners provided work space, the urban proleteriat provided labor, and the workers lived outside the factory gates. The separation of noisy, polluted industrial zones from leafy garden suburbs, and their linkage by commuter transportation networks, became one of the great triumphs of enlightened urban planning.

In a parallel development, telegraph and telephone networks enabled the spatial separation of management from industrial production and allowed managers and clerical workers to cluster with others of their kind in downtown central business districts. The focusing of transportation networks on these centers, and the concentration of network infrastructure there, drove up land values and encouraged high-density development. Steel, concrete, glass, and electrically powered systems provided the means, and the high-rise office tower emerged in response to these conditions; it became the characteristic business workplace of the twentieth century.

There were numerous variations on the theme of the coarse-grained industrial city, with its separated industrial, commercial, and residential zones linked by commuter transportation networks. You might get suburban office parks with parking lots instead of high-density, transit-serviced cores, secondary and tertiary centers, clusters of towers at edge-city locations, and vast, multicentered urban regions such as that of Los Angeles. But in major cities the principle of spatial separation into discrete zones, combined with that of commuter linkage among zones, obtained with remarkable consistency throughout the twentieth century.

As the capabilities of wired telecommunication networks developed and their service areas expanded, though, there was increasing speculation that telecommunication might substitute for travel within this sort of urban structure.[8] (If Dilbert was isolated in his cubicle, with his computer and his telephone, it might not much matter where that cubicle actually was.) One version of this idea was the concept of home telecommuting; the cubicle might become a study in a suburban house.[9] Another was that of the electronic cottage (Landor's Cottage with email); if you no longer had to commute to work, you might move your residence out beyond suburbia to a distant recreational or scenic area. Yet another version of the idea focused upon access to rural labor markets; if your business was in an expensive urban area where prospective employees could not afford to live, you might try to recruit a telecommuter workforce from less expensive rural areas. And another, again, was motivated by urban workspace costs; employers realized that they could save on rent by relocating employees to less expensive, suburban satellite locations instead of keeping them all at expensive downtown sites, with the added advantage that employees could save on commute time and costs.

But these strategies merely substituted one fixed workplace for another, and forced a tradeoff between them. As a result, they met with mixed success. There have been some interesting examples of live/work televillages in attractive locations, such as Richard Rogers's design for ParcBIT in the Balearic Islands,[10] and Giancarlo De Carlo's rehabilitation of the Ligurian mountain village of Colletta di Castelbianco,[11] and electronically serviced live/work has certainly played a role in the revitalization of urban areas such as Manhattan's SoHo, San Francisco's SoMa, and Singapore's China Square. Telecommuting has also proved to be crucial to rapid disaster recovery in the wake of events such as the 1994 Northridge earthquake in Southern California, and the September 11 attack on New York, which destroyed workplaces and disrupted commuter travel. For example, several thousand Lehman Brothers employees telecommuted in the weeks following September 11, while replacement work space was being found for them. However, by the early 2000s, it seemed clear that full-time telecommuting was not going to take off on a large scale.

The reason is simply that the tradeoffs don't necessarily look attractive unless there are some particularly compelling lifestyle, social, or economic motivations to consider. You might appreciate not having to commute, but you might also miss the companionship of coworkers. You might value the comforts of home, but you might also want opportunities to get out of the house. You might enjoy the view from your mountain hideaway, but you might discover that losing visibility at headquarters was a career disadvantage. You might like the flexible hours and childcare opportunities afforded by working at home, but you might realize that the home no longer served as a refuge *from* work. You might save on travel costs but incur the additional rent and operating costs of a home office. The workplace could fragment, and the home and the workplace could recombine—reinstating a preindustrial pattern in the postindustrial era—but this was not always and unambiguously to the good.

By contrast, wireless networks, portable electronic devices, and online work environments now allow information workers to move freely from location to location as needs, desires, and circumstances demand. By the early 2000s, many workers (and their employers) had discovered that they just needed a cellphone and a laptop to operate effectively at their nominal workplaces, on commuter trains, in airplane seats and airport lounges, in hotel rooms, at the sites of clients and collaborators, at home, and on vacation. Anyplace was now a potential workplace. And this condition would only intensify as the technology of nomadics developed and proliferated.

So the emerging, characteristic pattern of twenty-first-century work is not that of telecommuting, as many futurists had once confidently predicted; it is that of the mobile worker who appropriates multiple, diverse sites as workplaces.[12] As architects are rapidly discovering, this breaks down rigid functional distinctions among specialized spaces, and makes provision for varied and sometimes unpredictable functions increasingly critical; a home must serve as an occasional workplace, a hotel room must also be an office, a café table must accommodate laptops, and a workplace must adapt to more complex and dynamic patterns of use. Special places, with particularly desirable qualities, become powerful attractors when traditional person-to-workplace linkages are loosened; if you have your wireless connections, a seat under a tree in spring beats an interior office cubicle. And electronically arranged, ad hoc meeting places—where you can most conveniently form the human clusters you need at a particular moment, while remaining in wireless contact—dominate the fixed locations and inflexible schedules that had once been necessary to enable interaction and coordination.

HERTZIAN PUBLIC SPACE

Public spaces have traditionally worked best at nodes in transportation networks. The central crossroad is typically the focus of rural village life, and in larger settlements this crossroad often grows into a village green, a piazza, or a town square. Still more extensive cities—among them Rome, Georgian London, Haussmann's Paris, and Cerdá's Barcelona—have frequently been organized around multiple public spaces embedded in networks of streets and avenues.

These spaces become even more effective when they superimpose nodes of different networks. The village crossroad, for example, may not only play a role in local pedestrian circulation, it may also serve as a stop in a regional bus transportation network and the site of a public telephone box. A piazza may be activated by a well or fountain—that is, a public node in a water supply network. And multiple transportation networks may converge, as at Sydney's Circular Quay, where ferries, trains, buses, taxis, and urban pedestrians come together at a bustling, café-lined interchange and meeting point.

In many contexts, wired telecommunication networks have both taken advantage of the activity generated by this superimposition and added an important component to it. Central intersections and squares were natural sites for post and telegraph offices, and before private telephones were common and toll calls were inexpensive, these often expanded to provide banks of booths for long-distance phone service. Similarly, in the early Internet era, before network connections had become commonplace and computers had dropped to insignificant cost relative to income levels (a shift still under way in developing countries and areas), networked computer clusters emerged as new foci of public spaces—much like the traditional village well, but supplying a different type of scarce resource. At MIT, for example, Athena clusters unexpectedly developed into important meeting and socializing points. Public libraries also got connected and introduced computer clusters. Then, as the necessary infrastructure spread throughout urban areas, Internet cafés began to take root. It was not surprising that, when telecommunications infrastructure began to return to war-torn Afghanistan in 2002, one of the first points of Internet availability was an Internet café in the basement of Kabul's Intercontinental Hotel.

The hard-core Internet café, particularly in the developing world, was nothing more than a low-rent, fluorescently lit space jammed with as many computers as possible. In upscale versions, there was more space, some effort to create a pleasant ambience, and an overlay of the electronic action onto more traditional café functions; tables simultaneously accommodated (not always with complete success) food, drink, keyboards, and screens. Those that went for the fashionable crowd exploited the pervasive electronics to intensify the pleasures of seeing and being seen. New York's Remote Lounge, for example, provided cameras above each table to multicast to other tables—and, of course, to a Web site.[13] And as public screens became larger and more vivid, their role in creating outdoor spectacle grew— a gradual shift from café to carnival. Fields of LEDs began to take over and transform the roles of the movie poster, the theater marquee, the billboard, the cornice inscription, and even the entire urban facade. Shameless old Times Square, which had long been in the spectacle business, completed its rake's progress from a space demurely defined by traditional architectural elements, to one dominated by print and paint, then the glitter of lightbulb and neon, to programmed LED pixels; visitors encountered an retina-jangling spatial collage of animated graphics, electronic stock and news tickers, television programming (with closed captions instead of sound), reflexive imagery

(New Year's Eve crowds watching video images of themselves), and tourists getting their fifteen seconds of big-screen fame on Broadway.

Just as the social role of the village well began to fade with the introduction of domestic piped water supply, though, so did that of the public point of network presence as domestic space got wired and computers got cheaper. If you had a networked computer at home, you didn't need an Internet café, and the point of presence that really mattered to you was the switching node that connected your domestic line (the "last mile" of the network) to your service provider.[14] Public access points hung on longer in specialized contexts—where they provided significant technological advantage (in particular, faster and more reliable connection), as in India, where infrastructure is sparse and often unreliable and private connection remained beyond the reach of many, in China, where Internet cafés also served as connections to a wider world and rallying points for the intelligentsia, and in Korea,[15] where PC *baangs* allowed kids to escape from crowded domestic space and strict parental supervision and had a particular cultural resonance in the land of consumer electronic gadgets—but by the early 2000s it was clear that they would eventually shrink to niche roles at best. They would go the way of other hopeful space mutations, such as the telegraph station and the drive-in movie theater, which had emerged in response to particular technological conditions and were left behind by further technological change.

Meanwhile, continuous fields of network presence had begun to blanket public spaces. This began with cellular technology, which brought phones out of seclusion like bees from a hive and rapidly transformed public behavior and space use—most obviously by inserting the essentially private activity of phone conversation into space governed by the conventions of public conduct. This created frictions, which had to be resolved by the development of new rules of etiquette. Miss Manners had her work cut out for her: continuous phone accessibility was part of the point, but this produced rings at inappropriate times and places. You might call from a public place to avoid being overheard at home or in the office, but you might end up annoying strangers with your chatter; you might be so engrossed in your private conversation that you lost track of your immediate surroundings and walked into someone; your phone might inconveniently

ring while you were in company, forcing you to find a quick and grace-
ful way to excuse yourself; and, if you couldn't help overhearing an
embarrassing private conversation, you had to avert your eyes and
compose your face.

A more profound outcome was the new capacity of mobile
urbanites, at unknown locations, to arrange ad hoc meeting places.[16]
In the past, meetings had depended upon explicit prearrangement
(leaving you at a loss if, for some reason, your party didn't show up as
expected), upon random encounter, and upon standard meeting places
and regular schedules that increased the probability of random
encounter. Italian piazzas, for example, had traditionally depended for
their social efficacy upon their central locations and ingrained conventions of showing up there at established times; now, Italians simply
call one another and arrange meetings on the fly. Piazzas still look the
same and still work superbly as public space, but their patterns of use
have become far more flexible.

In dangerous urban contexts, such as the conflict-ridden cities
of the Middle East, continuous cellphone communication has become
crucial for tracking and intelligence. Parents use cellphones to keep in
contact with their mobile children, to check up on them when something happens, and to get up-to-date reports on current conditions. In
contested contexts, as when cities experience street demonstrations,
cellphones play an equally crucial coordination role; swarms of mobile,
wirelessly connected and coordinated demonstrators contend with
similarly equipped and coordinated squads of police. And, in less tense
situations, the same sorts of strategies drive the game of cellphone-
coordinated celebrity spotting.

With the emergence of 802.11 wireless networks in the early
2000s, a new field of functional possibility superimposed itself on
public space. This hub-and-card technology provided convenient connectivity for wireless laptops, and enthusiasts set about creating public
wireless hotspots. These first emerged in semipublic spaces—such as
cafés, bars, lobbies, waiting rooms, and airport lounges—which suddenly became much more useful as ad hoc workplaces and online interaction points; instead of reading a newspaper, you could download
your email or surf the Web. (If you wanted privacy, you learned to take
the seat against the wall, and keep the back of your screen to the

public.) Then hotspots migrated outdoors. Midtown Manhattan's Bryant Park was one of the first outdoor public places to provide for 802.11 surfing under the trees and email from a park bench. A latter-day Manet might paint laptop users on the grass.

In general, as these transformations of public space illustrate, there is a strong relationship between prevailing network structure and the distribution of activities over public and private places. Where essential urban networks have relatively few access points, as with public transportation and water supply networks, the access points are often in public places, attract activity to those places, and thus strengthen their roles as meeting and interaction points. When networks become more ramified—as with domestic water supply and sewer, automobile transportation, electrical, and communication networks—they tend to decentralize functions and move them into private spaces—the public bathhouse gives way to the private bathroom, the public theater to the private home entertainment center. And where networks go wireless, they mobilize activities that had been tied to fixed locations and open up ways of reactivating urban public space; the home entertainment center reemerges as the Walkman, the home telephone as the cellphone, and the home computer as the laptop.

VIRTUAL CAMPFIRES

In traditional nomadic societies, regularly rekindled campfires provided mobile focal points for social life. With urbanization, social life became more commonly focused by fixed attractions—village wells, domestic hearths, and computer network drops. In the mobile wireless era, a third alternative has emerged; we can use our portable communication devices to construct meeting points and gathering places on the fly—places that may only be known within particular, electronically linked groups, and which may only play such roles for fleeting moments.

10

AGAINST PROGRAM

As continuous fields of presence are overlaid on architectural and urban space, the ancient distinction between settlers and nomads—long the bedrock of our thinking about cities—is eroding in subtle but important ways. In the emerging wireless era, our buildings and urban environments need fewer specialized spaces built around sites of accumulation and resource availability and more versatile, hospitable, accommodating spaces that simply attract occupation and can serve diverse purposes as required. A café table can serve as a library reading room. A quiet place under a tree can become a design studio. A subway car can become a place for watching movies.

ELECTRONOMADIC SPATIAL PRACTICES

The relationships of mobile bodies to sedentary structures have loosened and destabilized; inhabitation is less about doing what some designer or manager explicitly intended in a space and more about imaginative, ad hoc appropriation for unanticipated purposes. We are becoming less like Saint Jerome, immobilized in his study among his accumulated possessions, less like Dilbert stuck at his computer in his cubicle, and more like cyborg foragers navigating through electronically mediated resource fields. We are relying less upon things (or people) being at fixed locations, or available on regular schedules, and more upon electronic tracking and navigation to locate what we want and take us to it. Our mental maps of buildings and cities are becoming less static records of fixed features and more dynamic representations of current conditions.

This condition, understood in the most optimistic way, offers liberation from the rigidities and interdictions of the predefined program and the zone—a release from ways of using spaces produced and enforced by dominant social orders.[1] It opens up the possibility of new, as yet unimagined spatial practices, and the opportunity (in the words of Michel de Certeau) "to rediscover, within an electronicized and computerized megalopolis, the 'art' of the hunters and rural folk of earlier days."[2] Or, if you don't like the pseudo-primitivism of this formulation, you might imagine rediscovering Baudelaire's *flânerie*,[3] situationist "drift,"[4] or whatever it was that Deleuze and Guattari were recommending in *A Thousand Plateaus*.[5]

Conversely, for those who would exert state or corporate power, it offers anonymity and the possibility of avoiding resistance. Today, such power may flow as easily from a fluidly and ambiguously located constellation of cellphones as it traditionally has from a throne room in a palace, a boardroom in a corporate headquarters, or a courtroom in a national capital. As resistance movements have quickly realized, sites for effective confrontation of power are becoming harder to identify.[6] How do you determine a time and locate a place for resistance? Where do you demonstrate? What do you occupy?

The evolution of taxi fleets has dramatized these transformations in the use and control of space. In the past, where urban densities were too low for drivers to rely upon customers hailing them in the street, centralized wireless dispatchers fielded telephone calls and assigned jobs. Now cabbies carry cellphones as well, and rely upon their mobile, distributed, peer-to-peer networks for intelligence about traffic conditions and tips about concentrations of potential customers. In more advanced systems, customers make location-coded cellphone calls, cabs have GPS navigation systems, and software assigns jobs based upon proximity. There is a shift from centralized coordination and control to electronically mediated swarming.

While their elders were trying to figure all this out, kids—employing the short text messaging capabilities of cellphones—imaginatively pioneered the new spatial tactics of ad hoc occupation and electronic appropriation. They quickly learned to fan out through city streets in fluid packs, electronically negotiating and specifying sites for assignations, raves, and street demonstrations. Those who wanted to repress these practices soon came up with the countermeasure (at least for the moment)—have the cops confiscate the phones. And the kids, in response, are discovering how to immobilize opponents by unleashing worms and viruses that clog channels of communication. Control of space—particularly in real time—now requires control of the airwaves.

In many ways, the dynamic ebbs and flows of the basketball court and the soccer field provide compelling models for these new spatial practices. The players are mobile, autonomous actors, but they are in constant visual and auditory communication with one another, and they adjust their actions in response to evolving situations. Over larger chunks of terrain, the wirelessly communicating units of a military operation act in similarly coordinated fashion. Now, spatially dispersed yet coordinated, fluid collections of wirelessly interconnecting individuals—perhaps assembled, from the beginning, in cyberspace rather than at any physical location—are becoming a crucial fact of urban life. They constitute a new category of human assemblage— one to add to our traditional conceptions of the gathering, the throng, the crowd, the masses, the mob, the cadre, the cell, the ensemble, the battalion, and the team.[7]

The connected masses also create problems of differential mobility. Traditional nomads understood these problems well and often dealt with them brutally; they left behind the aged, infirm, and otherwise immobilized. In the context of electronomadics, it is often a matter of relative reliance on bits and atoms, and the consequences tend to be economic. Scholars who can rely upon online resources are highly mobile and can work effectively on the road, but their colleagues who need access to undigitized print material or precious original manuscripts are still tied to traditional scholarly sites and practices. Telephone call centers can readily relocate and may want the flexibility to do so when it becomes economically advantageous, so they may be reluctant to invest in surrounding communities. Financial firms that had their premises destroyed in the World Trade Center attacks could instantly activate backup sites and send their employees into telecommuter mode, but restaurants and other small establishments that serviced those firms in Lower Manhattan were stuck at their sites, lost clientele, and suffered disproportionately badly. The new mobility divide may turn out to be more important than the digital divide.

THE DECLINE AND FALL OF THE ARCHITECTURAL PROGRAM

For architects, continuous fields of presence and the destabilization of person-to-place relationships demand some radical rethinking of the fundamentals. The standard procedure of twentieth-century modernism was to start by distinguishing and separating functions—the better to optimize spaces for particular functions and to announce those functions visually. (Communication engineers might think of it as space-division multiplexing of activities.) At an urban scale, housing areas were to be distinguished from industrial and commercial zones. At building scale, there were to be specialized spaces, with associated equipment, for the activities that were to be accommodated. And the physical fabric of a building was to be articulated functionally—for example, by separating the supporting and enclosing functions of a wall by substituting columns for support and a nonbearing curtain wall for enclosure. But this strategy makes little sense when wireless electronic devices can support many different activities at a single location or the same activity at many different locations, and when running different software can radically alter the functions provided by a device without changing its form at all. Time division multiplexing of activities is starting to look smarter than space division.

The key instrument of the traditional spatial organization strategy was the written *architectural program*—a detailed list of required spaces, specifying floor areas, technical requirements, and adjacency needs.[8] Built space made the provisions of the program concrete, and construction bureaucrats compared plans to checklists just to make sure. But the architecture of the twenty-first century can (if we choose to take the opportunity) be far less about responding to such rigid programs and much more about creating flexible, diverse, humane habitats for electronically supported nomadic occupation. It can be an architecture not of stable routines and spatial patterns, but, as Michael

Batty has suggested, of continually reconfiguring clusters of spatial events characterized by their duration, intensity, volatility, and location.[9]

This architecture will pursue the benefits of loose binding. Consider these in the context of office space, for example. When office workers have cubicles filled with files and bookshelves, it is relatively difficult and expensive to move them around; churn takes time and costs money, so managers have traditionally tried to minimize it—with the result that organizations are slow to adapt to change, and workers are often left in locations that no longer serve them well. But if the personal information environments of office workers automatically and instantaneously follow them around, they can sit down and work anywhere. The cost of regrouping to meet new needs drops almost to zero.

You can also look at this from a long-term space management perspective. When organizations move into new buildings, they usually have carefully worked out space plans. Then, over time, they make incremental changes in response to emerging demands, with the result that the space becomes fragmented and inefficient, much as the disk space does on your computer. Defragmentation is difficult and expensive when move costs are high, but it is easy when move costs become negligible. It is just like running Norton Utilities to clean up your disk.

Furthermore, physical enclosure for information security purposes now matters less, while electronic security matters more. It was once essential to ring cities with defensive walls, but that is irrelevant now. And it was once crucial to lock office doors, so that the papers and files inside could be protected against dispersal or destruction, and so that their confidentiality could be preserved. (In fact that was one of the main reasons for the very existence of private offices.) Espionage was a matter of clandestinely breaking in and stealing papers or making illicit copies. If your files are online, though, and accessible to you anywhere you log in, you do not have to work in a physically secure space. You *do* want to be sure that those files are regularly backed up and electronically protected against unauthorized access, and you probably want to sit so that your laptop screen is protected from prying eyes.[10] In other words, information security has been deterritorialized and shifted to a domain of abstract symbol manipulation.

Selectively (though certainly not universally), space-to-space relationships are loosening. For convenience and security, old-fashioned library reading rooms had to be adjacent to book stacks, but that constraint disappears when stacks become servers and carrels become wireless reception points. To make most efficient use of an expensive resource, office staff once needed convenient access to a central copying machine, but that imperative evaporates when making a copy becomes a matter of sending a file through a network rather than carrying an original to a machine, and when inexpensive, net-worked laser printers are widely distributed. As connectivity matters more, in many contexts, adjacency matters less, and architectural form is less tightly determined by the need to satisfy adjacency requirements.

Even established ideas of flexibility and adaptability require reconsideration. In the past, architects provided these qualities by introducing modular, demountable partitions and furniture, movable components, plug-in devices, and the like. Now the focus is shifting to self-configuring electronic environments—enabled by electronic devices that can immediately begin to communicate wirelessly with one another when they are brought into proximity and that can work together to support whatever activities are taking place.[11] Laptops are beginning to talk wirelessly to video projectors, projectors and cameras to printers, telephones to speaker systems, video cameras to monitors, PDAs to other PDAs, automobiles to gas pumps, and so on.

In some ways, then, we are returning to strategies and practices of preliterate, precapitalist times. Ancient Greek philosophers, for example, did not have offices and classrooms; they strolled with their students through the groves of academe. Then the Hellenistic Library of Alexandria became a site of immobile accumulation, the fixed focus of a unique community, and a place where scholars had to be. Today, the Web is our Library of Alexandria, and mobile wireless connection allows scholars to stroll once more—but without losing access to the resources they need. This does sit uneasily, of course, with some large, petrified chunks of the Western philosophical tradition. If you are a Heideggerian, you will probably fret about "wandering" versus "dwelling." And, if you take the Hegelian position that surrounding oneself with tangible property is a way of imprinting your presence on the world (Jerome's books did not just serve his needs, they defined him), then you will be dispirited by digital dematerialization and networked server access. Perhaps, though, this just means that giants of thought are still creatures of their time—and maybe, in these cases, too prone to generalize from the stability and clutter of the bourgeois drawing room.[12]

ELECTRONIC NON-PLAN

At a larger scale, the instrument for distinguishing and separating functions has long been land-use *zoning*. This sometimes has a commonsense and unobjectionable function, as for example in mandating the separation of residential areas from noxious industry. But it has frequently been used to enforce far less benign forms of segregation. And there are far fewer good reasons to separate activities—such as working, being entertained, and pursuing your social life—when they are all supported by the same wireless, portable devices, and when, unobtrusively handled in this way, they do not interfere with the activities of others. There is, then, a new kind of opportunity to recoup the "right to the city," which Henri Lefebvre powerfully characterized in terms of heterogeneity rather than monoculture, encounter rather than separation, and simultaneity instead of sequence, and which he saw as threatened by "discriminatory and segregative organization."[13] Land use planners might move toward Lefebvre's "diversification of space," in which "the (relative) importance attached to functional distinctions would disappear."

The sixties Anglo-American counterpart to Lefebvre's insistence on the right to the city was a provocative call for "non-plan," set forth in a notorious *New Society* article by Reyner Banham, Paul Barker, Peter Hall, and Cedric Price.[14] In it, the authors bluntly claimed that "the most rigorously planned cities—like Haussmann's and Napoleon III's Paris—have nearly always been the least democratic," and asked "What would happen if there were no plan? What would people prefer to do, if their choice were untrammeled?" This comported with contemporary architectural interest in combining serviced megastructures with plug-in and disposable architectural elements that could be configured by inhabitants themselves—theoretical propositions such as John Habraken's "supports,"[15] Yona Friedman's *architecture mobile*,[16] and Peter Cook's Plug-in City. It also resonated with more pragmatic architectural experimentation focused on flexible "mat" buildings,[17] extensible structures, and "long life, loose fit"[18] design strategies.

These proposals vividly expressed the possibility of flexibility and freedom of choice, but they mostly didn't deliver. Large-scale physical reconfiguration of architectural space in response to changing needs has remained a slow, cumbersome, and expensive process. Furthermore, occupiable space is still a scarce resource, and physical reconfigurability does little to diminish problems of space allocation and coordination. But the proponents of non-plan had glimpsed another possibility in what was then known as the "cybernetic revolution." They wrote: "The essence of the new situation is that we can master vastly greater amounts of information than was hitherto thought possible—information essentially about the effect of certain defined actions upon the operation of a system." Planning had depended upon "simple, rule-of-thumb value judgments" that were held to have "perpetual validity, like tablets of the law." Today, they concluded: "Physical planning, like anything else, should consist *at most* of setting up frameworks for decision, within which as much objective information as possible can be fitted." In other words, information infrastructure that provides a framework for dynamic decision making is more powerful than physical megastructure. If you want adaptability, responsive software beats reconfigurable hardware.

Several decades later, of course, the non-plan group's faith in "objective information" and "scientific management" seems uncritically naive. (The remaining members would, no doubt, be the first to say so.) But mobile connectivity, combined with reduced reliance upon immobile resources, has heightened the need, which they so presciently identified, to replace predetermined space programs and rigid plans with swiftly and sensitively responsive, electronically implemented space management strategies. By the early 2000s, we could see the beginnings of this in the combination of electronic road pricing and electronic navigation systems for managing road real estate, the combination of electronic tracking of parking space occupancy and automatic direction to vacant spaces, and flexible assignment of office cubicles to mobile, laptop-equipped workers. It is no longer the architectural programmer who controls space use, and thereby expresses power; it is now the software programmer.

EXTREME ELECTRONOMADICS

What if we could go all the way with shaking ourselves loose, shuck the last few atoms from our souls, and simply live on server farms somewhere? The gonzo endpoint of these trajectories of dematerialization and hypermobilization is the suggestion that mental life is just an affair of bits in the brain; you might strip them from this squishy substrate (much as one rips a CD) and download yourself onto disk. You are, on this view, just software—and as device-independent as a Java applet. You don't have to run on a high-maintenance meat machine. You no longer have to be, as Yeats so famously lamented, "fastened to a dying animal." Like saints and shamans in ecstasis, you loosen, to the ultimate, the binding of your persona to materiality and place.

Hans Moravec has speculatively described the necessary operation:

> Layer after layer, the brain is simulated, then excavated. Eventually your skull is empty, and the surgeon's hand rests deep in your brainstem. Though you have not lost consciousness, or even your train of thought, your mind has been removed from the brain and transferred to a machine.[19]

I'm not too sure about the brain science of all this; no doubt the inscription of information into organic neural networks is rather more complex that that of magnetic bits onto thinly spread iron oxide.[20] And I would be surprised (to say the least) if the continuity of personal identity turned out to be such a straightforward matter, or if the mind/body distinction reduced so neatly to software/hardware.[21] (Belief in this possibility is, of course, the extreme form of the digitalist dogma that "content" can always be cleanly separated from its current material embodiment.) But let us assume we can successfully read, decode, and copy all our brain files—the equivalents of WORD files of memorized text, JPG files of visual memory, MP3 files of unforgettable tunes, EXE files that specify how to get things done, and so on. Let us imagine a "postbiological future" in which "we will think of ourselves as software, not hardware."[22] What then?

It would put land use and transportation planners out of work; real estate requirements would now be measured in megabytes rather than square feet, mobility in terms of bits per second rather than miles per hour, and accessibility in terms of wireless network coverage. But the result is not disembodiment, in the sense of complete erasure of materiality. Nor is it reincarnation in humanoid avatar form. It is a more complex, spatially distributed, fluid, hybrid form of embodiment enacted with new hardware—one in which silicon, copper, and magnetic subsystems play a vastly increased role, while carbon-based subsystems play a diminished and no longer so privileged one.[23] Mortality reappears as a server crash. (There are some work-arounds, perhaps; you could implement reincarnation as restoration from backup, and transmigration of the soul as a hardware replacement strategy.)[24] So, why bother with the messy and problematic brain operation? By other means, anyway, we are already asymptotically approaching that networked cyborg state. Why insist on taking the carbon completely to zero?[25]

We are at the endgame of a process that began when our distant ancestors started to clothe themselves with second skins stripped from other creatures, to extend and harden their hands with simple tools and weapons, and to record information by scratching marks on surfaces. It picked up speed when our more recent forebears began to wire up telegraph, telephone, and packet-switching networks, to place calls, to log in, and to download dematerialized information to wireless portable devices. It is repeated whenever a child learns to do these things; for the cyborg, ontogeny recapitulates phylogeny. It is not that we have become posthuman in the wireless network era; since Neanderthal early-adopters first picked up sticks and stones, we have never been human.[26]

11

CYBORG AGONISTES

Palma Nova, near Venice, with its famous star-shaped fortifications, is a city of two tales.[1] You can read complementary narratives from the plan.

One tale is of enclosure. The walls, as in other ancient, medieval, and Renaissance cities, protected the concentrations of assets and sedentarized populations within from nomadic bandits and mobile armies without. In addition, as Lewis Mumford cogently argued in *The Culture of Cities*, the power of massed numbers in itself gave the city a superiority over thinly populated, widely scattered villages, and served as an incentive to further growth.[2] Density and defended walls provided safety, economic vitality, and long-term resilience. In extreme conditions, when the city was under siege, the gates were closed, the battlements were manned, and the city became self-contained for the duration. If you wanted to attack it, you called upon some technology to breach the defensive perimeter, like Joshua's trumpet, Achilles' wooden horse, Francesco di Giorgio's tunnel beneath the walls of Castel Nuovo, a battering ram, or a siege engine.

The second tale is of connection. The central piazza, surrounded by public buildings, is both the focus of the internal street network and the local hub of a road network that extends through the gates and out into the countryside, linking the city to others. The piazza is—like the server of a local ISP—a node at which nearby and larger communities are connected. When the gates are open, the city functions as a crossroads rather than a sealed enclosure, a place of interaction rather than one of exclusion.

Urban history is, from one perspective, a struggle of these narratives for dominance. Eventually, the network won. Mumford associated this victory with the rise of capitalism—a new constellation of economic forces that "favored expansion and dispersal in every direction, from overseas colonization to the building up of new industries, whose technological improvements simply canceled out all medieval restrictions." For emerging modern cities, "the demolition of their urban walls was both practical and symbolic."[3]

Superficially, modern Manhattan resembles a scaled-up version of Palma Nova; it has a regularized street grid, is surrounded by water, and can be accessed by a limited number of bridges and tunnels. But the networks are denser and more numerous, they extend the city's connections much further out into the world, and they provide many more functions. Road, rail, water, and air transportation links connect to local, regional, and global destinations. Water supply and sewer networks extend the island's hydrological footprint over a huge area and establish vital connections to distant collection, storage, and treatment sites. Mechanical air supply networks make the interiors of large buildings, and the city's many underground spaces, inhabitable. Pipes for hydrocarbons and wires for electricity densely blanket the built fabric with delivery nodes and extend the supply network to tank farms and power stations far out in the hinterland.[4] There is a wired telecommunications network that began with the telegraph, evolved into an analog telephone network, and is now transforming into a multifunctional digital system. And there are multiple forms of wireless networking—particularly broadcast radio and television, microwave, cellphone, pager, and 802.11. Horizontal network links make use of terrestrial, subsurface, and aerial real estate, while vertical links run through the service cores and chases of buildings.

Today, in the era of the network triumphant, the technologies of attack and defense have correspondingly transformed. You can still, of course, just obliterate a city with explosives and bulldozers, but there are now subtler means. You can bring down the networks it depends upon and just let it die. (This is not an entirely new strategy: in 537 C.E., the Goths cut the aqueducts serving Rome and forced its defenders to flee.) Even more effectively, you can hijack those networks and turn them back against their creators—delivering destruction by appropriating the very transfer and distribution capabilities that motivated their construction.

Furthermore, since digital networks increasingly control other networks, there is an entirely new type of threat to contend with. Not only may there be direct physical attacks against the "real property" components of critical urban networks, there may also be cyberattacks against the computers and networks that control those infrastructures. As a U.S. Congress committee reported in May 2002, a computer hack may cause "the same damage as a strategically placed bomb."[5] The same report added, "A cyberattack can originate from any part of the globe and from any nation, group or individual. The low cost of equipment, the readily available technology and cybertools, and the otherwise modest resources needed to mount a cyberattack makes it impossible for governments, much less businesses, to identify or track all potential cyberadversaries."[6]

CRASHING NETWORKS

The simplest way to crash a network is to block or sever a crucial link. Networks fail—sometimes catastrophically—when backhoes cut telephone or power cables, blood clots block arteries, or accidents jam the 405 Freeway. They are particularly vulnerable to this form of failure at points where there is no alternative route (or, at least, no efficient one) for traffic to take. That is why links such as the Khyber Pass, or San Francisco's Bay Bridge, are of such strategic importance; break them, and you disconnect large chunks of network, on either side, from each other.

It is even more effective, in general, to shut down a node—particularly one at which many links converge. If an earthquake destroys a freeway interchange, for example, travel in several directions is blocked at that point. And if you close a major airline hub, such as Chicago's O'Hare Airport, you produce a major air travel disruption. Furthermore, nodes often concentrate more crucial functionality than links. Sever a vein, and you may survive, but stop your heart and you're a goner. Yank a cable at the periphery of a LAN and most of the network continues to function, but crash the central server and the whole network goes down.

PROPAGATION OF FAILURE

If part of a network fails to perform its function, the trouble may propagate to other links and nodes—particularly where the network is tightly coupled. A sewer blockage can produce a backup to your sink or toilet, and a fender bender can quickly propagate traffic jams far back through a freeway network. This becomes particularly troublesome when there are no escape mechanisms, such as pressure valves and off-ramps; that's why New Yorkers dread getting stuck in the Holland Tunnel.

Even worse, propagation of overload conditions can cause progressive failure of network infrastructure. In an electric power network, for example, burnout of a transformer can propagate excess loads to other parts of the network, which fail in turn, and so on. It is much the same with nuclear chain reactions, and with progressive structural failures, like that of the World Trade Towers, where floors collapsed onto lower floors, which then became overloaded and collapsed themselves, and so on with gathering force. To be protected against progressive failure, networks need devices such as fuses or circuit breakers, which sacrifice themselves to prevent further propagation of damage.

In large, high-speed networks, such as modern power grids and the Internet, patterns of overload propagation and progressive failure are potentially very complex, hard to predict, and frustratingly difficult to control. Once they get started, even in a small way, there is an ever-present danger that they will grow explosively to produce large-scale, long-term damage. In 1998, for example, the power grid of Auckland, New Zealand, experienced a cascading failure that badly damaged all four of the central business district's main power arteries. The New Zealand Stock Exchange shut down until it could switch to backup power, the central business district was darkened for many weeks, air conditioners and refrigerators went out (in the Southern Hemisphere summer months), and the pubs had to serve warm beer—something the locals took particularly amiss.

In addition to hardware failure, software failure can also propagate alarmingly. In January 1990, a minor mechanical malfunction in an AT&T telephone-switching center in Lower Manhattan caused the switch to shut down and automatically reset its control software. When the switch came up again, it notified other switches around the country that they could begin routing calls to it. Unfortunately, these notifications triggered a control software bug, which shut down the switches that received it. When these switches restarted, they further propagated the problem, and so on. Before the problem could be diagnosed and corrected, it had affected all 114 switching centers in the AT&T system, severely disabled the system for nine hours, prevented 70 million of the 138 million long-distance and 800-number calls placed on the system that day from getting through, and caused hundreds of millions of dollars in business losses.[7]

And it can get even worse; since different types of networks are often functionally interdependent, failure of one type can produce failure of another. Telecommunication and electrical supply networks are particularly closely intertwined; telecommunication devices require electric power, and increasingly, power grids are managed by means of sophisticated telecommunication systems. Similarly, if the power grid goes out, the traffic lights cease to function, and the traffic network rapidly snarls up. And where pumps power water and air supply networks, power failure quickly renders many buildings uninhabitable. Even where there is no direct functional interdependency, the physical collocation of network links can propagate failures; for example, the tunnels into Manhattan provide both transportation and telecommunication conduits, so destruction would simultaneously affect both types of networks.[8] And the destruction of the World Trade Center towers took out both a major subway transportation node in the basement and a concentration of wireless telecommunication nodes on the roof.

One often-told, dramatic tale of cross-network cascading is that of a Worcester, Massachusetts, kid who hacked a telephone switch, wiping out custom settings and disabling telephone service in the area.[9] The automated control tower of the local airport used the telephone network to activate runway landing lights. Thus, when an approaching plane signaled the tower to switch on the lights, the lights failed to operate, and the airport had to be closed.

STRUCTURE AND VULNERABILITY

Sometimes, large networks fail—unsurprisingly—due to general destruction of their infrastructure. When an ice storm hit Quebec in

January 1998, for example, it brought down trees, poles, pylons, and tens of thousands of miles of power lines spread over a huge area. As a result, large parts of Montreal were left without power for most of a freezing month, and the emergency repair effort was enormous. But under some circumstances, highly localized failure or destruction can also produce major outages.

At the very dawn of the computer network era, pioneering researchers began to realize that the capacity of a network to continue functioning after damage depended a great deal upon its structure. Paul Baran introduced his seminal paper on distributed communications networks with a diagram showing three types of networks—centralized, decentralized, and distributed.[10] The centralized networks consisted of links radiating from a central node, exactly as with the radial roads of Palma Nova and other centrally planned cities. It was, as Baran noted, "obviously vulnerable as destruction of a single central node destroys communication between the end stations."

High-rise towers suffer from a similar vulnerability when, as in the World Trade Center towers, their vertical access routes are concentrated in central service cores. Building codes do require multiple, separated means of escape, and this may suffice in the case of relatively small fires, but it does not suffice when catastrophic events destroy cores entirely.

The decentralized network was a "set of stars connected in the form of a larger star," much like the patterns of major streets radiating from public places in the Wren plan for London, the Haussmann plan for Paris, and the L'Enfant plan for Washington. The centers of stars were still points of vulnerability, but the destruction of one such subcenter would not be a complete disaster; this would disconnect nodes directly linked to it, but the rest of the network could continue to function. Thus decentralization provided a way of scaling up and of isolating the effects of damage.

The distributed network was a nonhierarchical mesh, as in the street grids of Manhattan and Chicago, though not necessarily as regular. Its redundancy provided robustness; if nodes or links were destroyed, traffic could simply route around the damage. But there was an efficiency penalty; to get from one node to another, you typically had to pass through many intermediate nodes, which were potential points of inefficient transfer and congestion. If you travel through a street grid, you always have many options, and you are unlikely to get stuck if there's a blockage somewhere, but you will encounter many stop signs and traffic lights. Similarly, if high-rise towers were interconnected by sky bridges, they would enhance safety through the increased redundancy and wider spatial distribution of their escape networks, but they would be harder to control by means of security checkpoints at entry-level elevator lobbies.

In practice then, large transportation, communication, and other networks are often combinations of stars and meshes—seeking a balanced combination of the advantages of both. So Palma Nova has not only its radial arteries, but also several rings of streets concentrically around the central piazza, and smaller piazzas (forming mini-hubs) at the centers of its wedge-shaped segments. The radial streets of Paris and Washington are superimposed upon meshes and grids of more minor streets, and the radial transportation links of many modern cities are supplemented by ring roads and ring lines. Air transportation networks are increasingly organized around major hubs but continue to include links that do not pass through the hubs. And the vast structure of the Internet has turned out to be highly hierarchical, with a dominant pattern of major hubs, but also with significant amounts of meshing in many regions.[11]

ACCIDENT AND ATTACK

The large, decentralized networks that increasingly dominate our globalized world have turned out to be remarkably resistant to random accidents and failures. Nodes and links may go down here and there, but unless they unfortunately happen to be vital hubs or critical arteries, the effects are usually fairly brief and localized. But deliberate attack is another matter. Intelligent attackers can pick out the most attractive targets—going simultaneously for several major hubs, for example, rather than wasting their ammunition on peripheral nodes. Even a structure as large and heavily meshed as the Internet is probably very vulnerable to coordinated pinpoint attacks; there are good reasons to conceal, harden, and defend major switching nodes.

The attack on New York's World Trade Center towers vividly drove home these lessons about network structure. In the immediate aftermath, the surface street network, which is highly redundant, continued to function effectively. The New York subway system is much less redundant, so destruction of a major node beneath the towers did produce significant, long-term transportation outages. But the attraction of the target to the terrorists was not only the concentration of human life and the powerful symbolism of the towers, but also the role of Lower Manhattan as a key node in the global financial network—supported by an astonishing concentration of telecommunications infrastructure. In September 2001, there was more fiber-optic cable under the streets of Manhattan than in all of Africa, the two main telephone switches in the Financial District had more lines than many European nations, and there were more than 1,500 antenna structures on top of the World Trade Center north tower.

The destruction of telecommunications infrastructure was extensive. Verizon served more than three million local phone lines from its 140 West Street central office, which was badly damaged by debris, smoke, and water. AT&T had a central office in the basement of the World Trade Center; this survived the building collapse but lost electric supply and eventually went down when backup battery power gave out at 4 P.M. on September 11. It had served 20,000 T1 lines and 1,200 T3 lines to customers throughout Lower Manhattan and as far away as Long Island, so outages occurred not only locally but also in a random pattern throughout the region. At least fifteen cellular telephone base stations were lost, while many others had their landline connections knocked out by the damage to Verizon's facility. And of course the antennas on top of the north tower were all gone.

This loss of infrastructure combined with a surge in demand for telephone service, with the result that the telephone network became severely overloaded. On September 11, telephone traffic in the New York area was at about double the normal levels. Cellphone networks were jammed; during the morning, fewer than five percent of calls were connected. Nationally and internationally, AT&T connected 431 million calls—about twenty percent more than normal. To keep outgoing lines from New York and Washington open, AT&T blocked incoming calls. It required an enormous telecommunications recovery effort to get the New York Stock Exchange back in operation six days later.

The Internet continued to operate much as its designers had hoped.[12] There were some localized outages due to infrastructure damage and power failures, there was a surge in email traffic, and major news Web sites quickly became overloaded, but all this had little effect on the global Internet. However, the New York metropolitan area was, at that point, the Internet's largest single international bandwidth hub, and several of its major switching centers (carrier hotels) were all within close proximity on the west side of Lower Manhattan. It became apparent that a coordinated attack on the carrier hotels might have disconnected New York from the rest of the world, or the U.S. from Europe—though connections from the U.S. West Coast to Asia would still have continued to operate.

FEAR OF FOREIGN BODIES

But it can be better strategy to exploit, rather than destroy, the networks of your enemies. When you can obtain access to a large-scale network that they have obligingly constructed, you don't—like Napoleon marching to Moscow—have to expend a lot of effort moving your forces over large distances to reach your enemy. You don't even have to possess forces comparable to theirs. You can deliver violence and destruction, with modest means but pinpoint accuracy, by infiltrating or hijacking their networks.

The AIDS epidemic provided a brutal preview of this strategy. The HIV virus, as we quickly discovered, efficiently propagates itself through the network established by human sexual contact and by blood transfer—a network that has, in recent decades, been vastly extended through high-speed travel and population mobility. It has infiltrated itself into this human construction and has hijacked it for its own purposes. And in doing so, it has created a distributed state of siege—not the geographically focused sort, such as Palma Nova's walls were designed to withstand, but one manifested at a million condoms. Unlike the plagues of old, which required population density to flourish and could often be kept at bay by isolating and quarantining populations, AIDS depends upon the existence of connected paths—maybe extending over vast distances—within a global network.

The SARS epidemic of 2003 was a later variant on this theme. The virus was airborne, so face masks rather than condoms became the first line of defense. International gathering places—hotels, restaurants, air terminals, and airliner cabins—provided entry points into the long-distance air transportation network, which quickly propagated cases from the point of origin in China to Hong Kong and Singapore, then throughout the world. The more you relied on this network, the more exposed you were.

Computer viruses are chunks of mobile code in digital form, rather than genetic code in bioformat, but they turn out to operate in much the same way. They are too familiar, now, to require detailed explanation, and their potential for destruction has grown with the Internet. Just a week after September 11, for example, the Nimda virus struck 85,000 servers throughout the world, producing far more Internet congestion, outage, and economic damage than the infrastructure losses resulting from the World Trade Center attack.[13] Such viruses dramatically illustrate the downsides of decentralizing production and dematerializing functionality; they can be produced by even the moderately computer literate (right down to very modestly skilled script kiddies) at any of the world's millions of Internet nodes and propagated rapidly from those nodes. They can even be injected wirelessly into the Internet from mobile and transient locations. And, like the HIV virus, they generate a globally distributed state of siege—this time manifested at email filters to exclude suspicious incoming messages, virus-protected PCs, and corporate network firewalls.

Miniaturization, biotechnology, and nanotechnology provide yet more opportunities to infiltrate networks and turn them destructively back upon themselves. If you can produce small quantities of powerful toxins, virulent bioagents, or even vicious little nanobots, you can potentially distribute them to precisely where they will do most damage, through air and water utility networks.[14] Air-conditioning duct networks seem as if they were designed expressly for this purpose; they provide efficient conduits from conveniently accessible intakes to all the inhabited spaces of a building, bypassing guards and locked doors. In this case, the distributed siege points (now being put in place, in some contexts, as the danger is recognized) are filters, valves, and electronic sensors designed to sniff out and divert threats.

Finally, the miniaturization of destructive power has extended the possibility of infiltration to all forms of transportation networks. The Unabomber did not need B-52s or million-dollar missiles to deliver heavy bombs; he could use the mail system to put sufficiently deadly quantities of explosives right into the hands of his victims. For anthrax spores, the mailed packages can be even smaller. If you are prepared to be a suicide bomber, you can just drive a car, hail a cab, or take the bus to your destination. If you can get your hands on a van or truck, you can take a fertilizer bomb to a building's loading dock or underground garage, or simply drive it at the front door— hence the rows of bollards, Jersey barriers, and other ad hoc fortifications that now deface many urban landscapes. If you can compromise shipping containers, you can probably deliver nuclear weapons into the hearts of major cities.[15] And, particularly in the aftermath of the September 11 hijackings, airport check-in gates have become the most vivid and obtrusive reminders of the emerging state of distributed siege.

The densely, globally networked world is emphatically not (as early cyberspace utopians had sometimes imagined) inherently one of self-regulating, libertarian harmony. The proliferation and geographic distribution of access points—the very essence of the benefits of networks—also multiplies and distributes opportunities to create threats to the safety and well-being of those who have come to rely upon network capabilities. The entry barriers often aren't high; you can fashion many network-friendly destructive devices without much specialized knowledge, skill, or resources, and download or order much of what you need from the Internet. (As Martin Amis wrote in the immediate aftermath of September 11, "A score or so of Stanley knives produced two million tons of rubble.")[16] High-speed, efficient transfers within networks make these threats very difficult to localize and isolate, so that destructive effects may be felt far from the site of an initial security breach. And effective, inexpensive, widely deployable filters and barriers are not easy to devise.[17]

The characteristic fear of our times is no longer of the barbarians beyond the gates (or beyond the Cold War missile shield), but of

foreign bodies networked into our midst.[18] In the context of transportation networks and human movement, it manifests itself in the
understandable fear of the terrorist infiltrator and the suicide bomber,
and in the indefensible (sometimes overlapping) ugliness of anti-
immigrant, anti-refugee, and anti-minority demagoguery; the associated pathologies are closed borders, apartheid, and ethnic cleansing.
At another level, it is the fear that containers and vehicles—from
letters to airliners—will be infiltrated, hijacked, or redirected to serve
as carriers of explosives and toxins. Within communication networks,
it is the fear of viruses, worms, hackers, and crackers. And within networks of water and air transfer, and body-to-body contact, it is the fear
of deadly contagion.

DISPERSAL AND SWARMING

If you don't want to be an easy and perpetually anxious target in a
world of networks and distributed siege, one of your best strategies is
to decentralize. Instead of concentrating your business organization in
a conspicuous downtown office tower, you might disperse it to a collection of electronically interconnected suburban locations. Instead of
running your terrorist organization from a base that is subject to preemptive and retaliatory attack, you might create a network and scatter
its members throughout the community. In both cases, however, there
is something to lose; intercommunication becomes more difficult
and less effective, scale economies may disappear, and loss of regular
face-to-face contact may result in decline in trust and cohesion among
your members. But as the effectiveness of telecommunication increases
and the risks of centralization increase, the balance may shift in favor
of greater dispersal.

The events of September 11 drove home the lesson that centralizing an organization at one location, such as a skyscraper carrying the
corporate logo, does make it terribly vulnerable. As one New York real
estate manager commented in the wake of the World Trade Center
attack, "I'm not sure that tall trophy office buildings will ever be
popular again."[19] And as William Safire proposed in his op-ed column:
"We're a big, roomy country. Physically decentralized government,
tied together electronically, would be the strength of our nation if DC

were paralyzed."[20] These responses were probably exaggerated (particularly by anti-urban conservatives), but it was clear that the tradeoffs in the balance of centralization versus dispersal would henceforth be evaluated differently.[21]

Technically, the agglomeration economies that motivate the clustering of functions within organizations, and of businesses in industrial clusters, are offset by the risks of intense spatial concentration. Where the benefits of agglomeration remain high relative to the risks of natural disasters, localized network failures (such as power blackouts), or terrorist attacks, clusters are likely to remain. But, where the risks of agglomeration are higher relative to the benefits (and, perhaps, where these risks are reflected in insurance rates and taxes to support protective services—the so-called terrorism tax), and where efficient network interconnection reduces the advantages of agglomeration, organizations are more likely to spread risk through decentralization.

By taking advantage of wireless connection and miniaturized, portable equipment, you can add the strategic benefits of mobility to those of dispersal. Decentralization is by no means an entirely new idea among theorists of conflict and combatants. There have long been scattered, mobile guerrillas, irregulars, and resistance groups, but inexpensive, efficient, mobile telecommunication adds a new dimension to their operations; such groups can now act in much more effectively coordinated ways. In offensive mode, they can mount simultaneous attacks on key nodes in widely dispersed networks, such as the servers and switching hubs of a computer network, or the transformer stations of a power grid. They can converge on a target, in coordinated fashion, from many directions at once. And they can swarm—suddenly and unexpectedly materializing at some location in order to accomplish their goal, then rapidly dispersing to avoid containment or retaliation. Electronically coordinated street demonstrators at the 1999 World Trade Organization meeting in Seattle demonstrated the efficacy of swarming,[22] and the Critical Mass anti-car bicyclists in the San Francisco Bay Area have done even better at suddenly appearing out of nowhere to "cork" intersections and choke traffic.[23] "Swarm warfare" has become a fashionable topic among security analysts.[24]

The same technique could extend to telecommunication infrastructure. Even where nodes are multiplied and widely distributed, as with the Internet and cellular systems, they remain fixed targets. If nodes become wireless and mobile, however, the network itself transforms into a swarm that can rapidly reconfigure, elude attack, and move into areas of damage to restore service. After the 1994 Northridge earthquake, for example, cellphone providers developed a strategy of deploying trailer-mounted mobile cell sites, together with temporary, point-to-point microwave links to replace lost landlines and rapidly restore service in damaged zones. After September 11 in New York, this strategy was effectively used to reconstruct cellphone coverage and reweave it into the larger network. And military strategists have begun to contemplate swarms of collaborating, robotic aircraft interconnected by an "Internet in the sky."[25]

REPLICATION AND REPLACEABILITY

If dispersal, mobility, and elusiveness do not sufficiently reduce your vulnerability, you might turn to replication and replaceability. This is not a new idea either; generals have frequently been prepared to throw replaceable foot soldiers into the breach, automobiles have reserve tanks, and building codes require alternative fire exits in case one gets blocked. But the availability of cheap, plentiful electronic devices gives the idea a new spin. The Internet, for example, is built from relatively inexpensive channels and switches—which make extensive redundancy feasible and thus enable routing around damage sites. Even better, a mobile, ad hoc network, supported by portable, disposable wireless devices, would combine the advantages of fluidity and redundancy. This would be particularly difficult for enemies to take out as nodes could be scattered quickly around, they would not have fixed locations, and a lot of them would have to be found and eliminated to disable the network. If they did succeed in destroying some nodes, new ones would just pop up anyway.

Defensive strategies that rely upon redundancy work even better with software than with hardware, since software can quickly and inexpensively be replicated, and the process of replication can readily be decentralized. Thus file backup and distribution of backup copies to distant locations have become standard practice. Once, you backed up your PC by saving work to floppy disks or tapes; now, increasingly, backup is a network function. Whenever you send an email, for example, you create copies on multiple servers, and those servers are automatically backed up at regular intervals. When you work on a networked computer, it is increasingly difficult to assure, when you wish to do so, that a file you create is *not* getting backed up somewhere.

Organizations that rely upon digital data and cannot afford downtime frequently create redundant sites with both duplicate hardware and replicated data. Financial firms, in particular, have come to depend upon backup "hot sites"—with duplicate hardware, replicated software and data, and a crew of operators—that are ready to take over at a moment's notice. When the Lehman Brothers Manhattan data center was destroyed in the collapse of the World Trade Center towers, the chief technology officer was able to activate a backup facility in New Jersey, from his Blackberry text messenger, while escaping down the stairs. The firm was trading again the next day.[26] By noon on the eleventh, all major New York banks had activated their disaster recovery plans. Cantor-Fitzgerald, which lost 700 employees, was trading from backup centers in New Jersey and London when the bond market reopened two days later.[27] Many other affected financial firms used their disaster recovery contactors (SunGard, Comdisco, and others) to retrieve data from off-site backup locations.[28] But law firms, which depended more upon original documents on paper, recovered with much more difficulty. And at least one architecture firm, which did not have off-site backup, had to retrieve files piecemeal from the servers of its collaborators and consultants.

If instant recovery capacity is not necessary, backup sites can disperse over great distances. But where large amounts of data must be transferred to and from backup sites, and where high-speed recovery is imperative, network capacity limits dispersal. Thus, for example, the ESCON protocol used by IBM mainframes to connect to remote mass storage devices is limited to about a forty-kilometer radius. And the backup sites for Manhattan financial firms are mostly to be found elsewhere in Manhattan, in New Jersey, and in Brooklyn.

Since September 2001, however, there has been heightened interest in putting distance into distributed computing architectures, so that computer clusters can extend beyond potential disaster footprints. This requires combination of high-speed data links capable of operating over hundreds of kilometers with servers and storage systems designed to absorb increased load and to continue operating seamlessly when some of the nodes in the cluster go down. It seems likely that this sort of enterprise continuity technology will continue to develop.[29]

Switching to a backup site is not, of course, a possibility for the small lunch place adjacent to the large financial firm. In fact, the electronic relocation of its neighbor's center of activity may be disastrous for the lunch place, leaving it without customers. Strategies of electronic backup and redundancy are powerful, but their effects are differential.

SELF-REPLICATION AND MUTATION

You can shift this principle to a metalevel; the process of replication and dispersal can, itself, be replicated, dispersed, and mobilized. This makes it more robust than a process centralized at a single, potentially vulnerable production facility. Parasites, bacteria, and viruses provide the model for this; under favorable conditions they are self-replicating, and they can distribute themselves simultaneously through multiple channels, making them particularly difficult to stamp out.

Since the operation of writing information in memory is fundamental to digital computation, the logic of replication quickly manifested itself on early computers. Programmers learned to code loops that would write the same information repeatedly into memory—rapidly filling it to overflowing. With slightly trickier logic, and abandonment of the distinction between program and data (easy in languages like Lisp), they could write chunks of code that replicated themselves.

At this stage, the worst outcome of out-of-control replication was crashing an isolated computer—and you could fix the damage by rebooting. But the interconnection of computers into networks instantly changed that. It became possible to send destructive code from computer to computer—either explicitly, or more insidiously, by clandestinely attaching it to email or other transfers. Furthermore, it was quite trivial to write code that automatically replicated itself whenever it landed on a new host, then attached itself to outgoing email to propagate further. Thus in the Internet era computer viruses successfully imitated their biological predecessors.

Mobile, self-replicating code can be benign (like many biological viruses), and it can even provide an efficient way of performing useful functions on a large scale, but it can also wreak destruction. As Internet users have discovered to their chagrin, maliciously circulated viruses can overwrite memory, erase files, make software misbehave, display offensive messages, take over your machine to attack other machines, or simply crash your system. The varieties of damage that viruses can inflict are limited only by the imagination and technical skill (and it often doesn't take a lot) of programmers with bad intentions and network connections.

By now a standard (indeed, often essential) defense against viruses is software that scans incoming mail to detect and block attached viruses and also scans disks to detect and eliminate any viruses that may have lodged there. But the difficulties are just like those encountered with defenses against biological viruses—you need different remedies for different viruses, there is a continual threat of new viruses for which no defenses exist, and some viruses may replicate themselves with mutations that allow them to elude the defenses. Thus there is a problem of scale and complexity and an ongoing, escalating battle between virus and antivirus forces.

Code replication not only provides the means to propagate viruses, it also provides a way to amass forces for sudden, large-scale attacks from multiple directions. In distributed denial-of-service attacks on Internet servers, hackers surreptitiously take over many machines, then employ them simultaneously to fire streams of packets at target servers, thus overloading and bringing them down. Furthermore, denial-of-service attacks can be directed not just at a single server, but at multiple servers all at once, potentially providing a way to overcome the Internet's defensive redundancy. In October 2002, for example, a denial-of-service attack was directed against nine of the

Internet's thirteen root servers scattered around the globe.[30] A sustained, successful denial-of-service attack on all thirteen of the root servers would crash the Internet.

BEYOND INTERNET VIRUSES

By destroying digital resources, disabling the computers we have come to rely upon, and disrupting communications, Internet viruses can inflict immense economic damage, but they mostly aren't a direct threat to human life and safety. However, this will change as embedded networked devices proliferate, as our bodies become network nodes, and as transportation, power distribution, water, and air supply networks are increasingly intertwined with telecommunication networks. The price we will pay for integration and intelligent management of large-scale networks is that of continuously and resourcefully defending them against more and more potent virus threats.

Furthermore, as capacity for fabrication of physical artifacts shifts from centralized factories to small-scale, networked, personal fabrication facilities, traditional monopolies on production and distribution of weapons begin to break down. Personal fabrication of printed texts, toys, or electronic components may be unambiguously a good thing, but personal fabrication of guns and bombs, from downloaded designs and apparently innocuous materials electronically ordered from scattered suppliers, certainly is not. Personal biotechnology—maybe the fabrication of viruses from online genetic information and mail-order supplies—is even scarier. By 2002 researchers had, for example, fabricated infectious polioviruses using a publicly available genome sequence, inert chemicals, and modest laboratory equipment.[31] Self-replicating nanobots (which eliminate the distinction between production machinery and the products of that machinery) are not inconceivable—though many scientists are skeptical. And the most apocalyptic scenario is of a world suddenly overwhelmed by runaway, self-replicating gray goo. As Bill Joy has suggested, we now have the possibility "not just of weapons of mass destruction but of knowledge-enabled mass destruction (KMD), this destructiveness hugely amplified by the power of self-replication."[32]

Where cities from Troy to Palma Nova defended their encircling walls, New York and other twenty-first-century cities must defend their distributed networks against accident and attack. They must protect physical network infrastructure against destruction not only locally but also to far-flung extremities. They must ensure that there is sufficient redundancy in vital networks to prevent their being vulnerable to failure through destruction of a few key nodes or links. They must introduce circuit breakers, relief valves, and similar protections against failure propagation. They must find effective ways to guard against introduction of explosives, toxins, bioagents, portable code, and other destructive agents, and to guard against hijacking of vehicles, servers, and similar delivery devices. And cities must contend with both threats of physical destruction and threats to the logical integrity of networks from viruses, worms, software attack tools, and the like.

Conversely, if cities can keep their networks operating in times of disaster, they can quickly mobilize regenerative resources. Transportation networks can bring in relief supplies from distant parts of the globe. Mobile wireless nodes can swiftly restore telecommunications. And, increasingly, high-speed digital linkages to distant backup sites and geographically distributed enterprises can keep economic activity cranking along.

Traditionally, there was safety in numbers and in surrounding walls. Now urban security and resilience are grounded in patterns of connectivity. And defensive rings have fragmented and recombined. They no longer surround entire settlements and separate them from the countryside; rather, they enclose countless, scattered network access points—from airport departure gates to password-protected personal computers.

12

LOGIC PRISONS

According to popular lore (and numerous unfortunate jokes), Saint Peter stands guard at the pearly gates, downloads your lifetime track record when you show up looking for admission to the Heavenly City, and decides whether you get access privileges or not. Similarly, Cerberus provides the underground with security against unauthorized entry by the likes of Orpheus. Today, electronic control of movement among physical and online spaces works in much the same way. You had better be on the authorized user list if you want to get in to a secure Web site or access an ATM machine. And your profile had better not match that of the bad guys if you want to get on an airplane.

Technologies of surveillance and data collection, pattern recognition and data mining, and identity management have now converged with those of access management to enable formidable new systems of social control. At the core of these systems are access control *lists*, and their inverse, blacklists. Such lists include and exclude, distinguish friend from adversary, and in doing so construct political identity. To create them is to exercise power, and to be on them (or not) is to be subject to power.

ELECTRONIC ACCESS CONTROL

The idea of an access control list first arose, in its modern, digital form, in the early days of multiuser computing.[1] Users did not want others to be able to mess with their files, so operating systems were designed to keep track of the users who could read from, write to, or copy particular files. Essentially, operating systems maintained tables listing users along one axis, and files along the other, with entries specifying whether access was to be allowed or not.[2] A row listed the files a particular user could access, and a column listed the users who could access a particular file. And there were caste systems; the most privileged users got access to everything, the least privileged got very little.

Classification simplified this management task. If managers could formulate policies in terms of broad classes of users rather than individuals, and classes of things to be accessed, they could quickly reduce the sizes of access control tables and the complexity of associated management tables. They could establish and enforce general rules, for example, that only high-level staff within an organization could access files of salary data or enter the card-keyed executive dining room.

As large computer networks developed, access control systems became even more crucial. Since anyone might potentially access a site from anywhere in the network (anywhere in the world, in the case of the Internet), it was imperative to define, for each particular site, precisely who was allowed to do so; if you didn't, it was like leaving your front door unlocked to the whole world. In online communities, and persistent multiuser games—which might engage millions—the associated caste systems often evolved into elaborate structures that expressed themselves in terms of ability to change features of the shared environment. Lowly users might dress and drive their avatars, but more privileged ones might create persistent architecture, and the most privileged might alter the very infrastructure and rules of the game. Where code is the law, access to code makes you legislator and law enforcement.

Furthermore, malicious hackers have now become much more sophisticated and have figured out that that they do not need to rattle doorknobs electronically themselves; they can write code to search automatically through the Internet for vulnerable machines to crack. So access control has become not just a matter of managing human users, but also of fending off automated break-in attempts. The scale and complexity of the access management problem has rapidly grown.

Today, as more and more devices get embedded intelligence and IP addresses, access control lists are extending their power into the everyday physical world. They have accomplished a species jump, from cyberspace to architecture.

This has been prompted, in part, by the spread of embedded intelligence. If a home appliance has a processor and an IP address, it becomes potentially hackable (a Bulgarian teenager could reset your alarm clock, lock your air conditioner at a teeth-chattering temperature, turn all your audio devices up to maximum volume, and flash your living room lights on and off), so it needs access control. Thus your domestic access control lists begin to define who, physically or telepresently, is welcome in your house. The same goes for your automobile: instead of keys, future automobiles will probably have lists of authorized drivers.

In general, locks on gates and doors are evolving from metal-to-metal mechanisms to programmable electromechanical devices. They are increasingly being networked; and they are also getting lists of authorized users. (Instead of turning a key in a hotel room door, you swipe a card.) In other words, the technologies of access control in physical space and cyberspace—once very distinct—are converging. And so, of course, are the possibilities of misuse and abuse. Governments might control citizen movement by instantly reprogramming physical access to public spaces or by using electronic road pricing systems to create electronic cordons that are prohibitively expensive to cross, police in pursuit of fugitives might create instant holding cells around them, management engaged in labor disputes might program electronic lockouts, merchants might automatically exclude you from their stores if your credit isn't good enough, and burglars might simply reprogram your front door.

Furthermore, delivery of access control, like that of many other functions, can migrate from architectural elements to portable, wireless devices. Parolees, for example, may now be required to strap GPS-enabled tracking devices to their ankles.[3] Supervisors can establish exclusion and inclusion zones for each offender. According to the American Probation and Parole Association's manual on electronic offender supervision technology, "Exclusion zones are areas the offender is not permitted to go, such as parks and schools for a pedophile, a former partner's home or place of employment for a domestic batterer, or bars for an alcoholic. . . . Inclusion zones are areas the offender is expected to be at various times, such as his workplace during the day and home at night." The inclusion and exclusion zones are entered on a screen-displayed map, and the system transmits an alert "any time the offender enters an exclusion zone or leaves an inclusion zone at the wrong time." These systems can be extended to monitor for alcohol and blood consumption, and it is not difficult to imagine versions that could deliver painful electric shocks (a technique already used on dogs) or immobilizing drugs whenever a location or behavioral violation is detected.

But the technical details of electronic access control are not, in the end, the crucial issue; you can accomplish much the same results by putting the functionality into the architecture or attaching it to people. It is far more important to focus upon the specifics of access control policies in particular contexts—and on who has the authority to implement them.

IDENTIFICATION AND AUTHENTICATION

For good or ill, access control lists are simple and powerful, but they suffer from the obvious problem that names are not reliably unique; a glance at any large telephone directory or payroll file will confirm this. A related problem is that naming authorities may have overlapping jurisdictions and inconsistent policies. The remedy favored by armies, corporations, and some central governments is to issue everyone a unique serial number.[4] (The ancillary consequence is that, if you don't have a number, you're a nonperson.) This is a straightforward technical solution, but it may invest far too much power for comfort in the issuing organization.

An even more serious problem is that names may be stolen; someone seeking access may not be who he claims to be.[5] Generally, then, access control systems require you to verify your identity in some way. A standard authentication strategy is to request some piece of information that a random access seeker is not likely to possess, such as your mother's maiden name, and to check this against information that you have previously supplied. A now familiar variant, where better security is required, is to request a secret password—but, as most of us have discovered, it is easy to forget or confuse passwords, and it is easy to write software that will cycle through permutations of characters to find a password that works. A yet more sophisticated approach is to encrypt information and require a secret key to decrypt it. But even encryption, in the end, just shifts the problem to another level; just as you can steal a name, you can (usually, admittedly, with a bit more difficulty) steal a password or a key as well.

Instead of requesting information you carry in your head, an access management system can demand some physical thing you carry in your pocket. In the pre-electronic era, this might have taken the form of a letter of introduction (maybe with a wax seal), a passport, an ID card, or a metal key that you inserted into a matching lock— all of which, of course, might be lost, stolen, faked, or duplicated. Today it is likely to bear a barcode, a magnetic stripe, a memory chip that encodes digital information, or a mechanism that performs the cryptographic task of computing a one-way function.[6] Thus, for example, you access an ATM machine, a hotel room, or a card-keyed building by swiping a plastic card. With this strategy, you can end up carrying many different cards and other physical identifiers in your pockets, and you are in deep trouble if your identification tokens are lost or stolen.

Finally, an access management system might not be concerned with what people *possess*, but might rely on measuring what they *are*— for example, by automatically matching their fingerprints.[7] This biometric strategy requires technology for quickly and accurately measuring biological characteristics, storing profiles that identify individuals, and matching profiles of access seekers against stored profiles. Electronic fingerprinting devices, hand-geometry assayers, face recognition systems, and iris and retina scanners have all been employed for this purpose. Success so far has been mixed, but there is little reason to doubt that the technology will continue to improve. Short of having your hands chopped off or your eyeballs poked out, you cannot accidentally lose your biometric identifiers, and it is usually difficult and painful to remove them deliberately. (It's not like cutting up a credit card.) Still, you can fool biometric systems with masks, gloves, and prostheses, and a clever makeup artist might steal your face, which, unlike a credit card, you cannot have reissued.

None of these basic authentication strategies solves the problem of the attacker who somehow breaches the defenses and then (like the soldiers hidden in the Trojan horse) has free reign to cause damage. So some of the most advanced security systems now check credentials electronically at frequent intervals. To stay logged into a particularly sensitive computer system, for example, you may need a pocket wireless device that continually transmits random numbers that match numbers stored in the access management system. If the transmission falters, or the numbers do not match, you instantly get disconnected.

For access managers—from systems operators to border guards—effective combinations of these identification and authentication strategies are increasingly crucial to success. For users of buildings and systems, and citizens of cities, the capacity to electronically verify one's right to use a name—in multiple contexts and over extended time periods—is now essential. Without such capacity, in a digitally networked environment with electronically managed access, there can be no freedom of movement and action.

IDENTITY IN MULTIPLE CONTEXTS

Most of us, in our daily lives, need access to many different physical and online places. Consequently, we must carry numerous access devices and remember multiple IDs and passwords. These may need to be updated as we change employers or affiliations, and as organizations merge, move, or reorganize. And the difficulties can be complicated by the use of different names in different contexts—perhaps a full, formal name as shown on your birth certificate and passport, a nickname at your local bar, a writer's pseudonym, a stage name, a vanity license plate, and different avatars and handles in different online environments. Not only is this condition a potential source of confusion and error, it also introduces annoying and wasteful delays as we are forced to reidentify and reauthenticate ourselves when moving from place to place. As our identities become more complex and multifaceted, they increasingly require sophisticated management, and software is emerging to fill the need.

The Microsoft Passport system, for example, is intended for use by Microsoft customers, business partners, and affiliates. It allows you to "use one name and password to sign in to all .NET Passport-participating sites and services."[8] You can sign up to "store personal information in your .NET Passport profile and, if you choose, automatically share that information when you sign in so that participating sites can provide you with personalized services."

Similarly, the Liberty Alliance Project promises to "deliver and support a federated network identity solution for the Internet that enables single sign-on for consumers as well as business users in an open, federated way."[9] Unlike Microsoft Passport, it does not depend upon a single, central authority to maintain personal information. It proposes, instead, that it will enable businesses and users to manage their own data through the use of an open, interoperable standard for federated network identity. Federated spatial identity might one day be managed in much the same way—emptying your pockets of multiple keys and swipe cards.

Another approach, proposed by the National Electronic Commerce Coordinating Council, would make use of "clustered IDs."[10] This represents a different tradeoff between the efficiencies of centralization and the privacy and liberty protections of decentralization. It would encourage the development of multiple identity management systems operating across different collections of businesses, government departments, and so on, and would not require a single identifier that followed citizens everywhere.

TRACKING AND PROFILING

You can explicitly *sign up* for access to a physical or online site, you can be *put* on a list (maybe without your knowledge) by an access manager, or more insidiously, you can be put on a blacklist. Lists that emerge in these ways are not static; they are constructed and reconstructed over time, as access managers decide whom to include and exclude. To provide a basis for these decisions, managers collect and record data about individuals—for example, the sorts of data you provide on an employment, driver's license, credit card, passport, or visa application. Subsequently, they may add information about your activities and behavior, such as traffic violations, purchases and payments, or border crossings. The technology for this developed enormously during the 1990s, as Web activity generated unprecedented quantities of behavioral data and as electronic access control became an increasingly pressing issue.[11]

To track your activities and behavior, and hold you accountable for them, access managers must be able to identify you and follow your identity through time; that's why bank robbers and Mardi Gras revelers wear masks—to create "time-outs," during which they cannot be identified. (And that's why alibis are also important.) Conversely, electronic announcement of your identity is a form of unmasking and, as such, creates opportunities for tracking and profiling. You explicitly generate such opportunities whenever you assert and authenticate your identity by providing a password or swiping a card. And you implicitly produce them when you carry RFID devices: if cars have transponders, you can electronically check who's in your parking lot, and if students in a classroom have RFID tags, teachers can electronically take attendance. Some biometric systems, such as face recognizers, also make identity electronically observable. At the extreme, GPS-equipped offender supervision systems can create continuous traces of where the wearer has been, and when.

Your activities online already get tracked routinely, and in detail. Whenever you log into a site as an authorized user, the files you access and the times you access them will be recorded—and system administrators will be able to trace through the log to see who you are and what you have been up to. (To facilitate this, many Web sites require you to register.) On electronic commerce sites, the record of your transactions becomes a valuable marketing tool. Even if you just use a browser to enter a publicly accessible Web site, you will deposit more information about your identity in the site's log file than you may imagine, and your activities on the site will be tracked.[12] Everywhere in cyberspace, you leave electronic footprints. The point was vividly dramatized when, in April 2002, a Princeton admissions officer clandestinely entered a password-protected Yale admissions decision Web site. The Yale system administrators traced back the source of the intrusion, and it was soon all over the newspapers.[13]

As electronic identification is increasingly required for access to physical spaces as well as online ones, electronic tracking inexorably follows. It is often needed for billing purposes, and of course it may be put to other uses as well. Your monthly credit card bill may list the times and places at which your transponder-equipped car passed through checkpoints, just as your cellphone bill lists the locations of calls. Similarly, a workplace that is subdivided into card-key-accessed rooms can keep detailed track of when employees show up and leave, and where and when employees spend their time at work—and it had better not be too much in the lunchroom or the washroom! In this respect physical space increasingly resembles cyberspace.

DATA WAREHOUSES

It might logically seem that all these electronic activity records would remain associated with the sites and organizations that collect them, to be used only in those particular contexts. This has traditionally been the case; employers, retailers, educational institutions, credit agencies, license issuers, and immigration authorities have all maintained their separate databases, and it has not been easy to consolidate them. This provides a natural privacy safeguard; if you don't want your activities at some physical or online site to be logged, you can simply avoid it— much as celebrities try to protect their privacy by avoiding being seen in public places.

Another possibility is that databases will increasingly be consolidated in data warehouses—thus creating more comprehensive logs of user activities and reducing the likelihood that users can escape the net. (Advocates like to use the jargon of eliminating data "stovepipes"—implying that it is just about business efficiency.) Nationwide identity cards and numbers greatly facilitate warehousing. Such consolidation has generally been opposed by privacy advocates, and centralized, consolidated databases would provide attractive targets for break-in attempts and denial-of-service attacks. However, nationwide identification systems and large-scale data warehousing have had more advocates since September 2001—particularly in the United States.[14]

It is also possible that databases will not be formally consolidated, but that resourceful programmers who gain access to multiple sites (either legally or illegally) will be able to link an individual's disparate records to construct a single record. The more common elements (such as names, addresses, and ID numbers) there are in different databases, the easier this ad hoc correlation becomes.

It seems most likely that personal data management will eventually evolve into a complex, distributed function in which some data are collected and managed at the physical and online sites you visit, some are maintained personally (much as you may now maintain resumes and personal Web pages), and some are maintained by trusted third parties who are in the business of guaranteeing completeness and accuracy. There will be associated, potentially regulated practices of transferring data among repositories, and correlating and cross-checking data from different sources. One way or another, though, as an inhabitant of an electronically networked world, you not only have a verifiable name, you have extensive records of your background and past actions associated with that name. Your reputation depends upon these records, you may be held accountable for them in various ways, and you may end up negotiating the conditions under which these records are constructed, accessed, and interrelated.

Large-scale database consolidation is not just a matter of technical efficiency; it fundamentally changes the conditions of urban life. Until recently, personal privacy depended heavily upon scale effects; small, traditional villages mostly provided very little of it, but large, modern cities allowed the possibility of disappearing into the anonymous masses. It also depended upon the localization of knowledge; if you wanted to leave your sordid small-town past behind, you could run off to join the French Foreign Legion, or just move to L.A. Now, in the global village, electronic data collection, storage, correlation and analysis technologies, combined with high-speed electronic transfer of information, have radically shifted the natural balance of power in favor of snoopers and list makers. Unless rigorously enforced privacy policies intervene, computer power overcomes scale, and networks defeat distance.

Once access managers have databases resulting from surveillance or tracking, they can implement software that looks for matches to things that interest them.[15] If they capture facial images of passengers passing through an airport security checkpoint, for example, they can check for close matches to photographs of suspected terrorists. Algorithms that perform this task are statistically based, so it is necessary to set a threshold of similarity. If this threshold is too high, they can expect relatively few positive matches, and there is a high likelihood that actual terrorists will slip through unrecognized. If it is too low, they will get a large number of false alarms, and, like the boy who cried wolf, the system will lose credibility.

More abstractly, access managers can look for digital signatures in data about personal characteristics and behavior. This is like searching for a few unusually shaped straws in a large haystack, but sophisticated analysis techniques combined with massive computing power make it increasingly feasible. For example, the U.S. Transportation Security Agency has pursued development of a passenger profiling system known as CAPPS II.[16] Credit card companies have long analyzed patterns of card use in order to detect fraud. And financial institutions have begun to monitor and correlate customer activity and to report suspicious patterns to law enforcement and intelligence agencies.[17] Of course, it all depends upon the definitions of target signatures—on what particular access managers might mean by "dangerous" or "suspicious."

It is not always necessary to predefine patterns of interest. For decades, numerical taxonomists and cluster analysts have computed measures of similarity among digital descriptions of things (such as customer activity profiles) and have used these measures to construct similarity classes. Online retailers, in particular, have learned to sort customers into groups with similar buying habits and to use this information for pinpoint marketing—recommending books and CDs when you log on to a retail site, for example. This is harmless enough when you find yourself clustered with purchasers of, say, English romantic poetry, but you may attract unwelcome attention if you get grouped with purchasers of bomb-making manuals and extremist political tracts.

It is also possible to extract pertinent facts from a behavioral profile, add rules, and apply an inference engine to derive potentially telling conclusions. For example, a system might infer that a car was speeding from the time intervals between charges at turnpike tollbooths. Or it might infer that someone had made a plane journey from the short time intervals between cellphone calls from widely separated cities.

Precrime does not require mutants floating in vats, as in *Minority Report*—just a database, rules and profiles, inference engines, and data mining algorithms. And, if the conclusions they draw are not one hundred percent accurate, then that can always be dismissed as collateral damage—unfortunate, but justified by the end result.

SCIENTIA EST POTENTIA

In the past, defenders of places and populations could mostly depend upon more traditional strategies for distinguishing between friends and adversaries. The matter could often be settled, very simply, by relationship to physical boundaries; if you were inside the city walls, the default assumption was that you were one of us, but if you were outside, you were presumed to be one of them. Sometimes, as well, it could be determined by ethnic or gender markers, or by military uniforms—blue versus gray, red coats, red shirts, brown shirts, and black shirts. But in a networked world, foes are sometimes shadowy, spatially dispersed, and mobile.[18] You cannot look down at them from the ramparts, they don't have evident concentrations of fixed assets, and there aren't obvious targets to bomb. So governments that need to protect their citizens from them must first construct their identities: it doesn't help to declare war on an abstract noun such as "terrorism" unless there is some algorithm—increasingly likely to be one that is applied to electronic traces of activities—for enumerating and locating members of the class of terrorists. Otherwise, the *Onion*'s famous 9/11 story, "U.S. Vows to Defeat Whoever It Is We're at War With," becomes all too prescient.[19]

And governments are quickly getting it—though not necessarily getting it right. After September 11, 2001, the Pentagon's Defense Advanced Research Projects Agency (DARPA) set up an Information

Awareness Office headed by the scary spook John Poindexter.[20] The IAO's publicly announced plans were to build a vast warehouse of personal data about individuals and to apply data-mining and pattern-recognition techniques to sniff out suspicious patterns of activity and electronically identify enemies.[21] Somebody at the agency got carried away with Photoshop: the Web site shows a variant on the great seal of the United States (as depicted on dollar bills), with an eye peeking out of a pyramid to scan the entire globe—a panoptic diagram that would have left Jeremy Bentham gobsmacked and Michel Foucault grabbing for his deconstructor. Around it is inscribed the motto *Scientia est Potentia*—knowledge is power.[22]

Indeed. But without vigorous, critical scrutiny of such power, there will be no clear boundary between protective and repressive uses of electronic surveillance and access control technologies. Communication networks are meant to be liberating, but they can also be used to construct personalized logic prisons for all of us.

Logic prisons define zones of inclusion and exclusion in both cyberspace and physical space. They are built not from stones and bars, but from access management lists, software, and electronic gadgets ranging from RFID tags and GPS anklets to card-key access locks. The guards aren't in watchtowers. They rely upon networked surveillance and tracking systems, data warehouses, and pattern recognition and data mining systems. And they exert their power through denial of access, apprehension at electronically managed checkpoints, and, where necessary, electronic fingering for arrest, immobilization, or elimination.

Next time you want to go somewhere, think of your friendly access manager. He's making a list, and checking it twice—gonna find out who's naughty and nice. Unless we are prepared to resist it, the logic prison is coming to town.

EPILOGUE

As early explorers of the southern oceans approached wild, dark shores, they were sometimes startled by the sight of campfires beyond the breakers. In 1520 Ferdinand Magellan named Tierra del Fuego (land of fire) for the numerous blazes that he observed there. And in 1773, for similar reasons, Tobias Furneaux named the Bay of Fires on the northern coast of Tasmania.

If you walk the shores of the Bay of Fires today, you can still see the hints and traces of a tragically vanished nomadic people. There are ancient shell middens among the cliffs. At the freshwater lagoons behind the dunes, near to the feeding grounds of wallabies and kangaroos, the now-silent campsites remain inviting. The fires and smoke plumes that impressed Furneaux were, for the brief moments that they flared into warmth and light, the centers of mobile communities— places where people gathered, talked, cooked, slept, and cared for one another.

For such nomads, the most basic and obvious ethical principle— and one crucial to social cohesion—must have been that of reciprocity: take responsibility for the welfare of others in the group because you expect them to do the same for you.[1] The sense of collective enterprise and mutual obligation tempered self-interest. But application of the principle of reciprocity was strictly geographically bounded: it would be hard to argue that the moral obligations of group members extended far beyond the circle of warmth—to include, say, the citizens of London, or even the inhabitants of the Australian mainland on the far side of the strait to the north. The Tasmanian aborigines had been completely isolated for thousands of years. There was literally no connection to the outside world.[2] The inhabitants of the Bay of Fires did not know of these distant people and could take no actions that affected them.

NETWORKS OF RECIPROCITY

The navigators saw the world differently. They were engaged in constructing a network of sea transportation routes that was to blanket the globe and form increasingly important linkages among remotely situated people. And these linkages, as those who took the trouble to reflect upon the question began to realize, affected the extension of reciprocity. The expanding circle of contact, trade, and capacity to do good and harm might have suggested that mere distance no longer made a moral difference, and indeed the seafaring eighteenth century was that of the first declarations of universal human rights. But the Victorian moral philosopher Henry Sidgwick was to give an answer more typical of the emerging colonial era, as follows:

> We should all agree that each of us is bound to show kindness to his parents and spouse and children, and to other kinsmen in a less degree: and to those who have rendered services to him, and any others whom he may have admitted to his intimacy and called friends: and to neighbors and to fellow-countrymen more than others: and perhaps we may say to those of our own race more than to black or yellow men, and generally to human beings in proportion to their affinity to ourselves.[3]

Implicit in Sidgwick's argument is the idea that moral obligation to one's fellow human beings rapidly diminishes with distance. In Sidgwick's world, parents, spouse, and children were likely to be very close at hand—probably in the same house. Other relatives, neighbors, and friends formed a wider geographic circle—mostly in the same town. Fellow-countrymen constituted a wider circle still, and those of other countries were very distant: the black people of Tasmania were a voyage of many months away. Casual racism aside, Sidgwick's analysis made considerable practical sense. Social groups and communities were largely place-based, and webs of mutual aid attenuated as they spread out from their centers, so it was reasonable enough to argue that obligation did as well. You had many obligations to those with whom you interacted intensely, over extended periods, but fewer to those with whom your interactions were limited or blocked by the impedance (some would say the tyranny) of distance.

But the racism that Sidgwick embraced turned out, of course, to be far from harmless. The ever-extending network of sea routes had, by then, brought Europeans and Tasmanians into close contact. And the argument that native Tasmanians had "little affinity to ourselves" was being used by European settlers to justify their extermination. Change the material basis of ethical principles—in particular, technologies of interconnection and networking—and the consequences may be momentous.

SCALING COMMUNITY

It is easy to see how networks of mutual obligation work in small bands and settlements, where people recognize one another. You can tell your friends from your enemies, grudges and favors matter, and it is important to protect your reputation. But what happens when villages grow into large cities? Both Plato and Aristotle argued that the social glue breaks down at the point where citizens become strangers to one another, so communities should not scale up beyond that.

In his *Laws*, Plato suggested that the population of an ideal city-state should not exceed about 5,000 citizen farmers, plus their families and slaves, and a few resident aliens.[4] And, in his *Politics*, Aristotle remarked: "In order to give decisions on matters of justice, and for the purpose of distributing offices of merit, it is necessary that the citizens should know each other and know what kind of people they are."[5] After pointing out that excessive population also made it "easy for foreigners and aliens resident in the country to become possessed of citizenship," he summarized: "Here then we have ready to hand the best limits of a state: it must have the largest population consistent with catering for the needs of a self-sufficient life, but not so large that it cannot be easily surveyed." In other words, it had to be possible to know your fellow citizens through face-to-face contact, and to build up webs of mutual aid and obligation through that contact. The central agora, which provided a place for all the citizens to meet, was the material, architectural expression of this idea.

Ferdinand Tönnies, pioneering German sociologist and contemporary of Sidgwick, provided a more nuanced suggestion—that the principle of reciprocity began to operate in different ways as settlements grew. *Gemeinschaft* (community), in his memorable formulation, was characterized by "intimate, private and exclusive living together" in families, kinship and friendship groups, villages and neighborhoods. *Gesellschaft* (society), on the other hand, was the "artificial construction of an aggregate of human beings"—the big, impersonal city.[6] It is certainly possible to quarrel with some of the details and subtleties of this distinction, and subsequent sociologists and social theorists have hardly been hesitant to do so, but it seems to me that Tönnies got it roughly right. *Gemeinschaft* doesn't scale, but other mechanisms take over. Cities are not just larger than villages: they are materially, socially, and ethically different.

Through the close proximities they establish, through shared use of public space, through common dependence upon complex and vulnerable infrastructure, and through joint access to vital networks of limited capacity, the world's great cities have become zones of dense, inescapable interdependencies among their citizens—most of whom know very little of one another.[7] They therefore depend upon sufficient levels of communication, cooperation, and trust among strangers; if this breaks down, cities cease to function. As the beating victim Rodney King famously pleaded in 1992, with Los Angeles exploding in riot and flames around him, "Can't we all just get along?"

It is not that cities are domains free of difference and conflict: rather, they rely upon the capacities of their highly evolved cultures and institutions—of ballot boxes, legal systems, markets, civil society, critical discourse, and political activism—to domesticate hostility, to constructively manage differences and conflicts, and indeed to generate creative energy from them. In other words, they are not just physical constructions, nor simply facilitators of economic, social, and cultural activity: they are also, crucially if imperfectly, systems of ethically guided interaction contrived according to the Rodney King principle—to allow large numbers of people, who do not necessarily know one another, to get along.

ELECTRONICALLY EXPANDING CIRCLES

As astronauts orbit the earth today, concentrations of humanity appear to them as glowing hotspots of light, heat, and radio frequency emission—much like campfires on the shore, but many orders of magnitude larger, and extending over a wider band of the electromagnetic spectrum. Within these zones of our Planet of Fires, supported by miniaturized electronics, the new spatial, social, and cultural practices that I have described in these pages are rapidly taking hold.

This third-millennium pattern is the culmination of a centuries-long process of weaving, superimposing, and integrating different types of transportation, energy, and communication networks. Subjects, extended bodies, settlements, economies, and cultures can no longer effectively be separated by skins, walls, and frontiers. They have all become inextricably embedded in dense, large-scale webs of interdependence. The child in Boston is socially and culturally linked to his grandmother in Melbourne, the server farm in Palo Alto is economically coupled to the cubicle farm in Bangalore, the cave in Afghanistan threatens the skyscraper in New York. If an outbreak of SARS is not controlled in Hong Kong, the consequences are immediately felt in Toronto. Mobility and interconnectivity on massive scales scramble the neat units and hierarchies described by Henry Sidgwick. Our circles of interaction and mutual obligation cannot be limited to our campsites, immediate neighborhoods, cities, nation-states, or even networks of international trading partners; they are truly and inescapably global.

Just as a city is not simply a sprawling village, the networked Planet of Fires is not a global village. Nor is it a virtual city or an extended nation-state. Physically, spatially, and morally it is a thing of a new kind—differing as profoundly from the civic arrangements we have known as *Gesellschaft* does from *Gemeinschaft*.

In principle, we could try to return to a simpler era by limiting the scope of our interactions or by denying that they create moral obligations. Those are the strategies, at different scales and in different ways, of the survivalist, the xenophobe, the isolationist, and the unilateralist. But, as Marx repeatedly argued, humankind never, in the end, rejects more effective means to satisfy its material needs. Few of us will choose to give up our instruments of liberation from what he might have called (had he lived to see it) the idiocy of unconnected life—the narrowly constrained existence imposed by limits of locality, time, memory, and processing power.[8]

We cannot withdraw to unambiguous home territories behind sealed borders because, in the world we have now made, there are none. And, under this condition of ubiquitous, high-speed, unbounded interconnectivity, we surely have no remaining grounds to argue that the principle of reciprocity is extinguished by distance. We are all tied together by our networks—both materially and morally—like climbers on a rope. If we are to reap the benefits of our electronically expanded social, economic, and cultural circles without succumbing to their dangers, we must recognize that they actualize our common humanity.

GLIMPSING THE GLOBAL POLIS

For the privileged and powerful, this densely and inextricably interconnected world can be a dangerous and frightening place. It must be controlled through total surveillance, comprehensive access management, preemptive arrests and strikes, and electronically administered high-tech violence. If you are not with the global superpower in this, then you are against it. For the marginalized and alienated, by contrast, it can provide opportunities to turn networks against their makers through infiltration, subversion, hijacking, and terrorism.

For yet others, this world can—and must—provide new opportunities to scale up networks of cooperation and mutual obligation, build borderless communities, and find common cause. The worldwide antiwar demonstrations of 15 February 2003 provided a tantalizing glimpse of this emerging possibility.

As far as I can tell, the idea of the demonstrations crystallized at a series of international meetings. Activists discussed it at the European Social Forum in Florence in November 2002, at a follow-up meeting in Copenhagen on 15 December, and at the World Social Forum in Porto Alegre in January 2003. (The World Social Forum, in contrast to the World Economic Forum in Davos, takes a bottom-up rather than a top-down perspective on responding to globalization.) Of course, all these meetings had associated Web sites, and these published globally accessible calls to action. It was the cyberspace equivalent of nailing a declaration to the cathedral door.

Before long, other Web sites throughout the world were publishing details of local demonstrations, and portal sites like stopwar.org.uk were providing hyperlinked directories. Some newspaper sites, such as Guardian Unlimited, quickly got into the act. And the sites began spontaneously cross-linking to one another. This tenuous cyberspace construction soon exploded into a huge, densely linked, self-organized network. If you entered at any point, you could, within seconds, find details of planned demonstrations at any city that might interest you. And the web of cross-linkages provided a strong sense of global solidarity. The capacity for bottom-up, uncoordinated growth that had been built into the Internet and the World Wide Web, and that had made them so effective, had been harnessed for new political purposes.

On the ground, the demonstrations began in Melbourne, on 14 February, then moved west with the sun. Sydney and other Australian cities followed on 15 February. Many time zones later, the wave hit Europe—Istanbul, Rome, Berlin, Paris, and London kicking off in succession. It reached New York on a bitterly cold Saturday afternoon, then crossed the continent to Los Angeles and San Francisco.

There was nothing virtual, immaterial, or dispersed about the events themselves. Their point was to achieve as much mass and density as possible, and to occupy the most prominent public spaces. Generally, they succeeded; in many cities, the crowds were the largest in memory. People put their bodies directly on the line, just as the ancient Greeks had done in the agora. It was face-to-face, physical, and sweaty. But these crowds were not blind masses, limited as in the past by line of sight; they were electronically coordinated swarms. Cellphones were everywhere. There was constant electronic interchange about evolving conditions and developments, so that the flows through the streets could shift and adapt in response.

As the tsunami of assemblies rolled on, and as if to celebrate the centenary of Marconi's first intercontinental transmission, long-distance electronic telecommunication helped it gather energy. Television reports and news Web sites carried vivid pictures of the events that had already unfolded in earlier time zones, so that those starting out in the next city knew that they were part of a huge global drama. It has long been obvious that market blips can propagate rapidly around the world through electronic networks, producing local consequences at each place they reach. So, it turns out, can political momentum.

When the demonstrations of the fifteenth were over, the invisible but powerful infrastructure of Web sites and cross-linkages remained. Furthermore, it had been invested, through the physical engagement of millions of people, with a credibility and emotional intensity that it had not attained before. And it was instantly pressed into service for the next phase of the burgeoning campaign; just surf in, and you could find your local point to demonstrate when the bombing began. Or, in a switch to the cyberspace side, you could join the 26 February virtual march on Washington—a Web-coordinated blitz of phone calls, faxes, and emails to every Senate office in the country. The *New York Times*, which had largely ignored the emerging voices of dissent, commented in a front page article that "there may still be two superpowers on the planet: the United States and world public opinion."[9]

ETHICAL INTERCONNECTIVITY

The "death of distance" that prophets of telecommunication have repeatedly promised[10] does not, as it turns out, destroy the power of place: local cultures and advantages still matter. Nor does it reduce to the extension of consumer service areas—to the convenience of calling your mother from a foreign city, watching CNN from any hotel room, getting cash from any ATM, or surfing the Web from an isolated mountaintop. It means, instead, that zones of networked interdependence are now growing in rapid, unbounded fashion. As they inexorably fuse into a single global system, they confront us with the challenge of imagining and forming extended social aggregates that are sustained not by force and fear, but by the ancient principle of reciprocity applied in new spatial patterns and on unprecedented scales—networks of ethical interconnectivity that allow scattered, disparate strangers, in vast numbers, to get along peacefully and productively.

These new civic formations will be embedded in particular physical structures—as surely as the walled city of Athens, the concrete and steel cities of New York or London, or nation-states and empires held together by their transportation and utility infrastructures. They will have geographic shape, and will result from investments in specific places. But they will be spatially discontinuous, overlapping and intersecting, and messily asynchronous in their patterns of daily activity. And they will be defined not by circles of warmth, not by surrounding stone fortifications, nor even by the borders and boundaries drawn on today's political maps, but by the endless hum of electromagnetic vibrations.

NOTES

PROLOGUE

1. The nymph Calypso provided Ulysses with onboard navigation capability, which got him as far as Phaeacia. Samuel Butler's translation of the *Odyssey* (book 5) describes the journey as follows:

    > He never closed his eyes, but kept them fixed on the Pleiads, on late-setting Boötes, and on the Bear—which men also call the wain, and which turns round and round where it is, facing Orion, and alone never dipping into the stream of Oceanus—for Calypso had told him to keep this to his left. Days seven and ten did he sail over the seas, and on the eighteenth the dim outlines of the mountains on the nearest part of the Phaeacian coast appeared, rising like a shield on the horizon.

2. Wireless telegraph technology had been around since 1896, and Marconi and others had conducted increasingly ambitious experiments—including transmission of a brief signal across the Atlantic from Cornwall to Newfoundland in 1901. But this was the first practical demonstration of the capacity to transmit wireless telegrams over great distances. Marconi's first message was a greeting from President Theodore Roosevelt to King Edward VII, via Poldhu station in Cornwall. Edward quickly replied "in the name of the British Empire."

3. The *New York Times* eventually used cellphone calls and email messages to put together a detailed chronicle of the last hours of the towers. See Jim Dwyer, Eric Lipton, Kevin Flynn, James Glanz, and Ford Fessenden, "Fighting to Live as the Towers Died," *New York Times*, Sunday, 26 May 2002, pp. 1, 20–23.

4. In September 2002, the President's Critical Infrastructure Protection Board dramatized the point by noting, "A train derailed in a Baltimore tunnel and the Internet slowed in Chicago. A campfire in New Mexico damaged a gas pipeline and IT-related production halted in Silicon Valley. A satellite spun

out of control hundreds of miles above the Earth and affected bank customers could not use their ATMs." *The National Strategy to Secure Cyberspace*, draft, September 2002, <www.whitehouse.gov/pcipb/cyberstrategy-draft.pdf> (accessed December 2002), p. 44.

CHAPTER 1   BOUNDARIES/NETWORKS

1. The duality of enclosures and networks is more than just a metaphor; it is a basic fact of graph theory—the mathematical study of network structure. Consider a floor plan as a planar graph in which corners are nodes and walls are links. Construct the adjacency graph of the plan by locating a node within every enclosed room, plus the exterior zone, then representing room-to-room and room-to-exterior adjacencies by links. The adjacency graph is the dual of the floor plan graph, and vice versa. The circulation network, created by doorways through walls, is a subgraph of the adjacency graph. For more detailed, rigorous development of this point, see Lionel March and Christopher F. Earl, "Architectural Applications of Graph Theory," in Robin J. Wilson and Lowell W. Beineke, eds., *Applications of Graph Theory* (London: Academic Press, 1979), pp. 327–56.

2. The decline of the city wall is often dated from 1494, when Charles VIII of France first deployed horse-drawn artillery pieces in his invasion of northern Italy.

3. Gottfried Semper noticed that the German word for "garment" (*Gewand*) is very closely related to the word for "partition" (*Wand*). He developed an elaborate theory of the relationships among walls, textiles, and clothing in his two great theoretical works, *The Four Elements of Architecture* (1851) and *Style in the Technical and Tectonic Arts, or Practical Aesthetics* (1860–63). (See Harry Francis Mallgrave, *Gottfried Semper: Architect of the Nineteenth Century* [New Haven: Yale University Press, 1996.]) In his essay "Housing: New Look and New Outlook," in *Understanding Media: The Extensions of Man* (New York: McGraw-Hill, 1964), p. 123, Marshall McLuhan repackaged the point: "Clothing and housing, as extensions of the skin and heat-control mechanisms, are media of communication, first of all, in the sense that they shape and rearrange the patterns of human association and community." More recently, Vito Acconci has produced a series of provocative works exploring his contention that "First there is skin and bones, then clothing, then a chair and then housing." See Sarah Boxer, "Poet Turned Antic Architect Keeps Exploring Inner Space," *New York Times*, 12 September 2002, pp. F1, F5. And the discourse continues with Claudia Benthien, *Skin* (New York: Columbia University Press, 2002), and Ellen Lupton, Jennifer Tobias, Alicia Imperiale, Grace Jeffers, and Randi Mates, *Skin* (New York: Princeton Architectural Press, 2002).

4. Georg Simmel, "Bridge and Door," trans. Mark Ritter, *Theory, Culture, and Society* 11 (1994): 5–10.

5.   For discussions of the pervasiveness of networks, see Albert-László Barabási, *Linked: The New Science of Networks* (Cambridge, Mass.: Perseus, 2002), and Mark Buchanan, *Nexus: Small Worlds and the Groundbreaking Science of Networks* (New York: Norton, 2002).

6.   In *The Production of Space* (1974; English trans., Cambridge, Mass.: Blackwell, 1991), p. 38, Henri Lefebvre argued, "The spatial practice of a society secretes that society's space; it propounds and presupposes it, in a dialectical interaction; it produces it slowly and surely as it masters and appropriates it." Lefebvre's analysis is extraordinarily suggestive, but it shows little curiosity about the specific technologies of spatial production and even less about the effects of changes in those technologies. In *The Informational City: Information Technology, Economic Restructuring, and the Urban-Regional Process* (Cambridge, Mass.: Blackwell, 1989), p. 6, Manuel Castells extended Lefebvre's argument by identifying "the emergence of a *space of flows* which dominates the historically constructed space of places, as the logic of dominant organizations detaches itself from the social constraints of cultural identities and local societies through the powerful medium of information technologies." In *Empire* (Cambridge: Harvard University Press, 2000), Michael Hardt and Antonio Negri proposed that the "irresistible and irreversible globalization of economic and cultural exchanges" had produced "a *decentered* and *deterritorializing* apparatus of rule that progressively incorporates the entire global realm," and "manages hybrid identities, flexible hierarchies, and plural exchanges through modulating networks of command." In this volume I shall be particularly concerned with the technological infrastructure of the global space of flows, the secretion of spatial patterns by means of that infrastructure, and the specific changes that are resulting from the development of a pervasive, wireless computation and telecommunication infrastructure.

7.   President's Critical Infrastructure Protection Board, *The National Strategy to Secure Cyberspace*, draft, September 2002, <www.whitehouse.gov/pcipb/cyberstrategy-draft.pdf> (accessed December 2002), p. 3.

8.   For a concise history of timekeeping technologies and their increasing precision, see William J. H. Andrewes, "A Chronicle of Timekeeping," *Scientific American* 287, no. 3 (September 2002): 76–85.

9.   Lewis Mumford, *Technics and Civilization* (New York: Harcourt Brace Jovanovich, 1963). See also Edward P. Thompson, "Time, Work-Discipline, and Industrial Capitalism," in Anthony Giddens and David Held, eds., *Classes, Power, and Conflict* (Berkeley: University of California Press, 1982), pp. 299–309; and David S. Landes, *Revolution in Time: Clocks and the Making of the Modern World*, rev. ed. (Cambridge: Harvard University Press, 2000).

10.   In his speculations on the evolution of machines, Samuel Butler suggested that large clocks would go the way of the big lizards. "Examine the beautiful structure of the little animal, watch the intelligent play of the minute members which compose it; yet this little creature is but a development of the cumbrous clocks of the thirteenth century—it is no deterioration from

them. The day may come when clocks, which certainly at the present day are not diminishing in bulk, may be entirely superseded by the use of watches, in which case clocks will become extinct like the earlier saurians, while the watch (whose tendency has for some years been rather to decrease in size than the contrary) will remain the only existing type of an extinct race." (Samuel Butler, "Darwin among the Machines," in *A first Year in Canterbury Settlement and Other Early Essays* [London: Jonathan Cape, 1923], p. 210.) Butler could not anticipate, of course, that tiny vibrating quartz crystals would one day make mechanical watches seem like bulky and expensive dinosaurs.

11. Ivan E. Sutherland and Jo Ebergen, "Computers without Clocks," *Scientific American* 287, no. 2 (August 2002): 62–69.

12. See David J. Bolter, *Turing's Man: Western Culture in the Computer Age* (Chapel Hill: University of North Carolina Press, 1984), and James Gleick, *Faster: The Acceleration of Just About Everything* (New York: Vintage Books, 1999), for discussions of accelerating subdivision and pace in the digital era.

13. In his October 1998 *Computerworld* column, for example, Edward Yourdon asked: "What if Y2K leads to massive corporate bankruptcies, heralding a long-term economic recession/depression? What if it leads to breakdowns in international communications, or a shut-down of the world's airports for six months?" This was over the top, and Y2K eventually passed with little incident. But the potential for some significant level of Y2K disruption had been real enough, and it had only been averted through a massive effort to identify and eliminate Y2K bugs.

14. Edward Yourdon, "What Comes after 1/1/00," *Computerworld* 32, no. 42 (1998): 89.

15. In 1884, at the International Meridian Conference in Washington, D.C., the globe was subdivided into twenty-four time zones, and the Royal Observatory at Greenwich was chosen as the prime meridian.

16. David P. Anderson and John Kubiatowicz, "The Worldwide Computer," *Scientific American* 286, no. 3 (March 2002): 40–47.

17. Seth Lloyd, "Quantum Mechanical Computers," *Scientific American* 273 (October 1995): 140–45; Seth Lloyd, "Quantum Computing: Computation from Geometry," *Science* 292 (2001): 1669; and George Johnson, *A Shortcut through Time* (New York: Knopf, 2003).

18. Carlton M. Caves, "A Tale of Two Cities," *Science* 282 (1998): 637–38.

19. Charles W. Moore, "Plug It In, Rameses, and See if It Lights Up, Because We Aren't Going to Keep It Unless It Works," *Perspecta*, no. 11 (1967): 32–43. Reprinted in *You Have to Pay for the Public Life: Selected Essays of Charles W. Moore* (Cambridge: MIT Press, 2001), pp. 151–60.

20. Mark Granovetter, "The Strength of Weak Ties," *American Journal of Sociology* 78 (1973): 1360–80.

21. Traditional, place-based communities were described, in many cases idealized, and contrasted with life in the big city in some of the landmark works

of sociology. See, in particular, Emile Durkheim, *The Division of Labor in Society* (1893; New York: Free Press, 1964), Ferdinand Tönnies, *Community and Society* (1887; East Lansing: Michigan State University Press, 1957), and Louis Wirth, "Urbanism as a Way of Life," *American Journal of Sociology* 44 (1938): 3–24.

22. Barry Wellman, *Networks in the Global Village* (Boulder Colo: Westview Press, 1999).

23. There is a growing empirical literature on the role of electronic interconnections in sustaining (or weakening) social networks. See, for example, Keith Hampton, "Living the Wired Life in the Wired Suburb" (Ph.D. diss., University of Toronto, 2001); Philip E. Howard, Lee Rainie, and Steve Jones, "Days and Nights on the Internet: The Impact of a Diffusing Technology," *American Behavioral Scientist* 45, no. 3 (2001): 383–404; Robert Kraut, Vicki Lundmark, Sara Kiesler, Tridas Mukopadhyay, and William Scherlis, "Internet Paradox: A Social Technology That Reduces Social Involvement and Psychological Well-Being," *American Psychologist* 53, no. 9 (1998): 1017–31; and Norman Nie, "Sociability, Interpersonal Relations, and the Internet: Reconciling Conflicting Findings," *American Behavioral Scientist* 45, no. 3 (2001): 420–35.

24. Barry Wellman, "Designing the Internet for a Networked Society," *Communications of the ACM* 45, no. 5 (May 2002): 91–96.

CHAPTER 2 CONNECTING CREATURES

1. In his 1865 essay "Lucubratio Ebria," Samuel Butler developed the idea that modern machines were extended limbs, and that these extensions evolved rapidly. "Every century the change in man's physical status is greater and greater. . . . Were it not for this constant change in our physical powers, which our mechanical limbs have brought about, man would have long since apparently attained his limit of possibility; he would be a creature of as much fixity as the ants and bees; he would still have advanced, but no faster than other animals advance." (Samuel Butler, *A First Year in Canterbury Settlement and Other Early Essays* [London: Jonathan Cape, 1923], p. 217.) Furthermore, he argued, command over such limbs was a privilege of class and power: "He alone possesses the full complement of limbs who stands at the summit of opulence, and we may assert with strictly scientific accuracy that the Rothschilds are the most astonishing organisms that the world has ever yet seen. For to the nerves or tissues, or whatever it be that answers to the helm of a rich man's desires, there is a whole army of limbs seen and unseen attachable; he may be reckoned by his horse-power, by the number of foot-pounds which he has enough money to set in motion. . . . Henceforth, then, instead of saying that a man is hard up, let us say that his organization is at a low ebb, or, if we wish him well, let us hope that he will grow plenty of limbs." ("Lucubratio Ebria," pp. 219–220.)

2. Thomas P. Hughes, *Networks of Power: Electrification in Western Society, 1880–1930* (Baltimore: Johns Hopkins University Press, 1983).

3. This process was to be exhaustively documented in the immediate postwar years by Sigfried Giedion in *Mechanization Takes Command* (Oxford: Oxford University Press, 1948). Giedion worried, as he meticulously enumerated wave upon wave of inventions, about "how far mechanization corresponds with and to what extent it contradicts the unalterable laws of human nature."

4. Telerobotic surgery, which had been a research topic for some years, became a practical reality on 7 September 2001, when a surgeon in a Manhattan office performed a gallbladder operation on a patient in Strasbourg. See Jacques Marescaux, Joel LeRoy, Michel Gagner, Francesco Rubino, Didier Mutter, Michel Vix, Steven E. Butner, and Michelle K. Smith, "Transatlantic Robot-Assisted Telesurgery," *Nature* 413 (2001): 379–80.

5. The Telegarden is a telerobotic art installation, located in the Ars Electronica Center in Austria, where Internet-connected remote users may plant seeds and water the flowers. See telegarden.aec.at and Ken Goldberg, ed., *The Robot in the Garden: Telerobotics and Telepistemology in the Age of the Internet* (Cambridge: MIT Press, 2000).

6. The idea of wireless remote detonation emerged in the very early days of wireless technology; by 1899 Nevil Maskelyne had demonstrated that gunpowder could be exploded by wireless remote control. When terrorist bombs exploded in Kuta, Bali, in October 2002, it emerged that SMS text messages to cellphones strapped to the bomb parcels had probably been used to detonate them. See Wayne Miller and Darren Goodsir, "Wanted: Police Name the Bali Six," *Age* (Melbourne), 18 November 2002, <www.theage.com.au/articles/2002/11/17/1037490052933.html> (accessed December 2002).

7. Manuel de Landa, *War in the Age of Intelligent Machines* (New York: Zone Books, 1991); James Der Derian, *Virtuous War: Mapping the Military-Industrial-Media-Entertainment Network* (Boulder, Colo.: Westview, 2001).

8. Norbert Wiener, *Cybernetics: or, Control and Communication in the Animal and the Machine* (New York: J. Wiley, 1948). In recent military discourse, this point has been articulated as the Revolution in Military Affairs doctrine. See Nicholas Lemann, "Dreaming about War," *New Yorker* (16 July 2001): 32–38.

9. For an overview of "the metabolic requirements of a city" and systems to serve them, see Abel Wolman, "The Metabolism of Cities," *Scientific American* 213, no. 3 (1965): 179–90.

10. The Biosphere was originally constructed, in the 1980s, to explore the possibility of long-term human survival in a closed environment. In the late 1990s, Nicholas Grimshaw's Eden Ecological Center, in Cornwall, was a less doctrinaire, more architecturally sophisticated exploration of the same theme.

11. Fredrik Liljeros, Christofer R. Edling, Luis A. Nunes Amaral, H. Eugene Stanley, and Yvonne Aberg, "The Web of Human Sexual Contacts," *Nature* 411 (2001): 907–8.

12. Dog-to-dog transfusions were attempted in the seventeenth century, following William Harvey's analysis of the functions of the heart and blood circulation system, and experiments with animal-to-human transfusions followed. In 1908 the French surgeon Alexis Carrel sutured an artery of a human donor to a vein of a recipient. By World War I, blood storage technologies had emerged, and military blood depots had been created.

13. For an excellent, detailed description, see Craig C. Freudenrich, "How Spacesuits Work," *Marshall Brain's Howstuffworks*, <http://www.howstuffworks.com/space-suit.htm/> (accessed December 2002).

14. Extracorporeal life support (heart/lung) systems generally include catheters to hook you up, connective tubing, a blood pump, an artificial lung, a heat exchanger, and monitoring devices.

15. The classic analysis of this move is Reyner Banham, *The Architecture of the Well-Tempered Environment* (Chicago: University of Chicago Press, 1969). For a French, post-'68 theoretical reading, see Dominique Laporte, *History of Shit* (Cambridge: MIT Press, 2000). And for valorization of the sanitary engineer and an argument that indoor plumbing turns the home into an "inseparable part of the urban body: the individual organ (the home network) becomes a member of the social body (the city's public network), so that the continuous link between private systems and equipment, and public network induces the shaping of a socially rooted and individually introjected image and model," see Alberto Abriani, "Dal sifone alla città," *Casabella*, no. 542/543 (1988): 24–29 (English summary, "From the Syphon to the City," p. 117).

16. Stephen Graham and Simon Marvin, *Splintering Urbanism: Networked Infrastructures, Technological Mobilities, and the Urban Condition* (London: Routledge, 2001); Matthew Gandy, "Water, Space, and Power," in *Concrete and Clay: Reworking Nature in New York City* (Cambridge: MIT Press, 2002), pp. 19–76.

17. For a retrospective view of Archigram, see Peter Cook, ed., *Archigram* (New York: Princeton Architectural Press, 1999).

18. In a retrospective discussion of the Plug-in City (Cook, *Archigram*, p. 39), Peter Cook commented: "The Plug-in City is set up by applying a large scale network structure, containing access ways and essential services, to any terrain. Into this network are placed units which cater for all needs."

19. Illustrations of these were published in Kircher's *Musurgia universalis* (Rome, 1650). For a recent discussion, which locates them in the history of surveillance systems, see Dorte Zbikowski, "The Listening Ear: Phenomena of Acoustic Surveillance," in Thomas Y. Levin, Ursula Frohne, and Peter Weibel, eds., *CTRL {SPACE}: Rhetorics of Surveillance from Bentham to Big Brother* (Cambridge: MIT Press, 2002), pp. 32–49.

20. The Singapore term "hand phone" comes closer to expressing this quality than the North American "cellphone." And it raises the question of how far *I* extend—to the tips of the fingers that grasp the phone, to the transmitting antenna, or to the receiving earpiece? The Greek term *kinito soma* also makes the association with the body (*soma*) explicit.

21. Thomas J. Campanella, "Eden by Wire: Webcameras and the Telepresent Landscape," in Ken Goldberg, ed., *The Robot in the Garden* (Cambridge: MIT Press, 2000), pp. 22–46.

22. The Japanese mobile carrier NTT Do Co Mo introduced G3 cellphone service in October 2001. The new G3 phones transferred data at about forty times the rate of earlier systems, crossing the threshold of feasibility for wireless videophones.

23. Michel Foucault, "Panopticism," in *Discipline and Punish*, trans. Alan Sheridan (London: Penguin, 1977), pp. 195–228. See also Gilles Deleuze, "Postscript on the Societies of Control," *October* 59 (1992): 3–8; and Levin, Frohne, and Weibel, eds., *CTRL {SPACE}: Rhetorics of Surveillance.*

24. By the spring of 2002, the inexpensive X10 wireless video camera had become a popular consumer item, and had opened up the possibility of drive-by electronic peeping by intercepting its 802.11b signals. See John Schwartz, "Nanny-Cam May Leave a Home Exposed," *New York Times*, 14 April 2002, pp. 1, 27.

25. Lou Hirsh, "Wireless Camera Takes Fantastic Voyage," *Wireless NewsFactor*, 11 January 2002, <wireless.newsfactor.com>. For these tiny, video-toting tourists, the intestinal tract is an amusement park ride, and the rest of the body is just mysterious poché.

26. The biorobot research team at Tokyo University, directed by Isao Shimoyama, has experimented with electronically controlled cockroaches, carrying tiny backpacks, for this purpose.

27. Details on these systems mostly have to be gleaned from investigative jour-nalists and the Web pages of watchdog organizations. See, for example, Duncan Campbell, "Inside Echelon: The History, Structure, and Function of the Global Surveillance System Known as Echelon," in Levin, Frohne, and Weibel, eds., *CTRL {SPACE}: Rhetorics of Surveillance*, pp. 158–69.

28. "Washington Plans Unprecedented Camera Network," Reuters, 13 February 2002.

29. There's a nod, in this, to Orwell, who also set his *1984* surveillance society several decades in the future.

30. For a lively analysis of this condition, together with some provocative argu-ments that general transparency is not necessarily a bad thing, see David Brin, *The Transparent Society: Will Technology Force Us to Choose Between Privacy and Freedom?* (Reading, Mass.: Perseus Books, 1998).

31. For a snapshot, circa late 1990s, see Brian Hayes, "The Infrastructure of the Information Infrastructure," *American Scientist* 85, no. 3 (May–June 1997): 214–18.

32. Adrian J. Hooke, "The Interplanetary Internet," *Communications of the Associ-ation for Computing Machinery* 44, no. 9 (September 2001): 38–40.

33. "Is it a fact . . . that, by means of electricity, the world of matter has become a great nerve, vibrating thousands of miles in a breathless point of time?

Rather, the round globe is a vast head, a brain, instinct with intelligence!" Nathaniel Hawthorne, *The House of the Seven Gables* (1851).

34. John C. Baker, Kevin M. O'Connell, and Ray A. Williamson, eds., *Commercial Observation Satellites: At the Leading Edge of Global Transparency* (Santa Monica: RAND Corporation, 2001), p. 1.

35. National Academy of Sciences, *Embedded Everywhere: A Research Agenda for Networked Systems of Embedded Computers* (Washington, D.C., National Academy Press, 2001), p. x.

36. Erin Hayes, "Detecting Terror: Lab Develops New Ways to Identify and Fight Terrorist Attacks," ABC News, 16 December 2002, <http://abcnews.go.com/sections/wnt/DailyNews/antiterror_technology021216.html> (accessed December 2002).

37. Paul Saffo, "Sensors: The Next Wave of Infotech Innovation," *1997 Ten-Year Forecast* (Menlo Park, Calif.: Institute for the Future, 1997). See also *Sensors*, <www.sensorsmag.com> (accessed December 2002), for coverage of emerging sensor technology. For a 2003 update see Gregory T. Huang, "Casting the Wireless Sensor Net," *Technology Review*, July/August 2003, <www.technologyreview.com/articles/huang0703.asp> (accessed June 2003).

38. The skin metaphor is becoming increasingly prevalent. See, for example, Neil Gross, "The Earth Will Don an Electronic Skin," *Business Week*, 30 August 1999, pp. 68–70.

39. In *The Consequences of Modernity* (Palo Alto: Stanford University Press, 1990), Anthony Giddens develops the argument that the institutions of modernity are characterized by such de-localization.

40. For detailed developments of this theme, see Mitchell Resnick, *Turtles, Termites, and Traffic Jams: Explorations in Massively Parallel Microworlds* (Cambridge: MIT Press, 1999); and Eric Bonabeau, Marco Dorigo, and Guy Theraulaz, *Swarm Intelligence: From Natural to Artificial Systems* (New York: Oxford University Press, 2000). For an attempt to apply concepts of feedback, self-organization, and emergence to urban design, see Steven Johnson, *Emergence: The Connected Lives of Ants, Brains, Cities, and Software* (New York: Scribner, 2001).

41. For snapshots of this phenomenon, as it manifested itself in the early 2000s, see Joel Garreau, "Cell Biology: Like the Bee, This Evolving Species Buzzes and Swarms," *Washington Post*, 31 July 2002, p. C1, and Howard Rheingold, *Smart Mobs: The Next Social Revolution* (Cambridge, Mass.: Perseus Books, 2002).

42. For extended development of the idea of markets as computational mechanisms, see Philip Mirowski, *Machine Dreams: Economics Becomes a Cyborg Science* (Cambridge: Cambridge University Press, 2001).

43. For a detailed discussion of Slashdot's feedback mechanism, see Johnson, *Emergence*, pp. 152–62.

44. The identification of emergent online communities has become an academic cottage industry. See, for example, Gary William Flake, Steve Lawrence, and C. Lee Giles, "Efficient Identification of Web Communities," in *Proceedings*

of the Sixth International Conference on Knowledge Discovery and Data Mining (Boston: ACM Special Interest Group on Knowledge Discovery and Data Mining, 2000), pp. 156–60.

45. J. C. R. Licklider, "Man-Computer Symbiosis," *IRE Transactions on Human Factors in Electronics*, HFE-1 (March 1960): 4–11. See also J. C. R. Licklider and Robert W. Taylor, "The Computer as a Communication Device," *Science and Technology*, no. 76 (April 1968): 21–31.

46. In the 1960s, Joseph Weizenbaum's Eliza system demonstrated that a system did not need deep intelligence to look smart in simple online conversation. By the late 1990s, Richard Wallace's ALICE chatbot could fool most of the people most of the time. It wasn't an intelligent entity, but it played one on screen.

47. Mary Catherine Bateson, foreword to Gregory Bateson, *Steps to an Ecology of Mind* (Chicago: University of Chicago Press, 2000), p. xi.

48. Bateson, *Steps to an Ecology of Mind*, p. 466.

49. Ibid., p. 465.

50. H. G. Wells, "World Brain: The Idea of a Permanent World Encyclopedia," contribution to the *Encyclopédie Français*, 1937, reprinted in H. G. Wells, *World Brain* (New York: Doubleday Doran, 1938).

51. <www.archive.org> (accessed December 2002).

52. On the dynamics of the Web, see Bernardo A. Huberman, *The Laws of the Web: Patterns in the Ecology of Information* (Cambridge: MIT Press, 2001).

53. The Internet Archive's collection is at <web.archive.org/collections/sep11.html> (accessed December 2002). See also <911digitalarchive.org> (accessed December 2002), and Genaro C. Armas, "Sites Archive 9/11 Communications," Associated Press, 19 August 2002.

54. Stephen Wolfram, *A New Kind of Science* (Wolfram Media, 2002).

55. Seth Lloyd, "Computational Capacity of the Universe," *Physical Review Letters* 88, 237901 (2002).

56. These arguments are frequently grounded in the philosophical positions of Heidegger and Merleau-Ponty. See, for example, Hubert L. Dreyfus and Stuart E. Dreyfus, *Mind over Machine: The Power of Human Intuition and Expertise in the Era of the Computer* (New York: Free Press, 1986); Hubert L. Dreyfus, *On the Internet* (London: Routledge, 2001); and George Lakoff and Mark Johnson, *Philosophy in the Flesh: The Embodied Mind and Its Challenge to Western Thought* (New York: Basic Books, 1999).

57. National Academy of Sciences, *Embedded Everywhere*.

58. Marvin Minsky, *The Society of Mind* (New York; Simon and Schuster, 1988).

CHAPTER 3 WIRELESS BIPEDS

1. Marshall McLuhan, "Housing: New Look and New Outlook," in *Understanding Media: The Extensions of Man* (New York: McGraw-Hill, 1964), pp. 123–30.

2. Peter Henderson, "No Office? No Desk! No Problem, Sun Says," Reuters, 1 June 2002.

3. In "Lucubratio Ebria," Samuel Butler drew a vivid contrast. "By the institutions and state of science under which a man is born it is determined whether he shall have the limbs of an Australian savage or those of a nineteenth-century Englishman. The former is supplemented with little except a rug and a javelin; the latter varies his physique with the changes of the season, with age, and with advancing or decreasing wealth. If it is wet he is furnished with an organ which is called an umbrella, and which seems designed for the purpose of protecting either his clothes or his lungs from the injurious effects of rain. His watch is of more importance to him than a good deal of his hair, or at any rate than of his whiskers; besides this he carries a knife and generally a pencil case. His memory goes in a pocket-book. He grows more complex as he becomes older and he will then be seen with a pair of spectacles, perhaps also with false teeth and a wig; but, if he is a really well-developed specimen of the race, he will be furnished with a large box upon wheels, two horses, and a coachman." (Samuel Butler, *A First Year in Canterbury Settlement and Other Early Essays* [London: Jonathan Cape, 1923], p. 218.)

4. The most notorious example is that of Burke and Wills, who set out to cross Australia from north to south with six camels and food supplies for three months. They violently drove off desert aborigines who offered them food and eventually died of starvation and thirst.

5. For an obsessively detailed discussion of the tradeoffs involved, see Colin Fletcher and Chip Rawlins, *The Complete Walker IV* (New York: Knopf, 2002).

6. In *Ulysses*, James Joyce reflected upon the path of water to Leopold Bloom's faucet:

   Did it flow?
   Yes. From Roundwood reservoir in county Wicklow of a cubic capacity of 2,400 million gallons, percolating through a subterranean aqueduct of filter mains of single and double pipeage constructed at an initial plant cost of £5 per linear yard by way of the Dargle, Rathdown, Glen of the Downs and Callowhill to the 26 acre reservoir at Stillorgan, a distance of 22 statute miles, and thence, through a system of relieving tanks, by a gradient of 250 feet to the city boundary at Eustace bridge, upper Leeson street . . .

   James Joyce, *Ulysses* (New York: Everyman's Library, 1997), pp. 902–3.

7. Chris Kenyon, "The Evolution of Web-Caching Markets," *Computer* 34, no. 11 (November 2001): 128–30.

8. For some recent explorations of this theme, see Jennifer Siegal, ed., *Mobile: The Art of Portable Architecture* (New York: Princeton Architectural Press, 2002).

9. Peter Cook, ed., *Archigram* (New York: Princeton Architectural Press, 1999), p. 119.

10. On the early history of wireless communication, see Hugh G. J. Aitken, *Syntony and Spark: The Origin of Radio*, 2d ed. (Princeton: Princeton University Press, 1985).

11. William Crookes, "Some Possibilities of Electricity," *Fortnightly Review* 51 (February 1892): 173–81.

12. "The World in Your Pocket," *Economist*, 7 October 1999.

13. William Stallings, "The Global Cellular Network," in *Wireless Communications and Networking* (Upper Saddle River: N.J.: Prentice Hall, 2001), p. 4.

14. The process has not been without its problems and setbacks. In the 1990s, the WAP wireless protocol failed to attract support, and Iridium low-altitude satellite phone service failed to attract customers. Providers and would-be providers of wireless services paid very high auction prices for spectrum licenses, and then had difficulty making their business models work. The 3G standard for advanced services, which was supposed to carry the industry forward to a new level in the 2000s, was not greeted with universal enthusiasm and found itself challenged by IEEE 802.11b (wireless Ethernet, or Wi-Fi) and Hiperlan. Still, from the early 1990s onward, in most parts of the world, the numbers of wireless users kept growing rapidly.

15. <http://grouper.ieee.org/groups/802> (accessed December 2002).

16. <www.bluetooth.com> (accessed December 2002).

17. <www.aetherwire.com> (accessed December 2002).

18. Very roughly, the essentials are as follows. Since antenna characteristics are basically the same whether an antenna is transmitting or receiving, the same antenna can be used for both transmission and reception. An antenna may be omnidirectional or (as, for example, with a parabolic reflective antenna) directional. The more powerful the signal radiated from an antenna in a particular direction, the further it will intelligibly carry in that direction. (Antennas can be very simple, but antenna design has evolved into a specialized technical area, and so-called smart antennas have emerged as an important new technology.) Signals with frequencies up to about 2 MHz travel by ground wave propagation; thus they follow the earth's curvature. Signals from 2 to 30 MHz travel by sky wave propagation; they bounce off the ionosphere, and can travel very long distances through sequences of hops. Above 30 MHz, neither ground wave nor sky wave propagation operates, so VHF, UHF, SHF, and even higher-frequency transmissions (all the way up to infrared and visible light) must rely upon line-of-sight propagation. For a comprehensive introduction to the technical issues, see Stallings, *Wireless Communications and Networking*. For a concise snapshot of emerging developments around 2002, see "Wireless Telecoms: Four Disruptive Technologies," *Economist*, 20 June 2002.

19. Development of Wi-Fi technologies began in the mid-1980s, when researchers at NCR, Apple, and other companies began experimenting with

wireless networks that employed unlicensed spectrum. The eventual outcome, in 1997, was the release of the IEEE 802.11b standard. Variants on 802.11, designed to serve different needs, soon followed. By the end of the 1990s, Apple was incorporating inexpensive 802.11 capability in its laptops. The technology grew in popularity in the early 2000s, and marketers and journalists began to refer to it as Wi-Fi. By 2003, Intel was marketing laptop chips with built-in Wi-Fi capability, *Wired* (*Unwired*, Special Wired Report, April 2003), and *Business Week* (28 April 2003) were putting out breathless special issues on "the Wi-Fi revolution," and the number of users had grown into the tens of millions. For a brief history, see Olga Kharif, "Paving the Airways for Wi-Fi," *BusinessWeek Online*, 1 April 2003.

20. These systems, in their initial form, had some significant technical limitations—particularly in their support of user registration, security, and mobility across base station service areas. However, it seemed likely that additional capabilities could be added incrementally as usage and demand for more sophisticated capabilities grew.

21. On community networking enabled by inexpensive client cards and 802.11b base stations, see Rob Flickenger, *Building Wireless Community Networks* (Sebastopol, Calif.: O'Reilly, 2002).

22. According to GrameenPhone's Web site, <www.grameenphone.com> (accessed December 2002), "In collaboration with Grameen Bank, which provides micro-credit only to the rural poor, GP utilizes the bank borrowers to retail telecom services in the rural areas . . . By bringing electronic connectivity to rural Bangladesh, GrameenPhone is bringing the digital revolution to the doorsteps of the rural poor and unconnected . . . Thus, the telephone becomes a weapon against poverty."

23. Licensing policies, in many parts of the world, have made this resource even scarcer by allocating broadcasters more spectrum than technically necessary. This made practical sense in the early days of wireless technology, but with newer technology it is now possible to make more efficient use of available spectrum, so the practice is less defensible.

24. The military sector, in particular, has often insisted on retaining control of licensed spectrum. See, for example, John Markoff, "Limits Sought on Wireless Internet Access," *New York Times*, 17 December 2002, <www.nytimes.com/2002/12/17/technology/17wire.html> (accessed December 2002).

25. For a concise summary of the technical and economic tradeoffs among different types of systems, see "Satellite," in National Research Council, *Broadband: Bringing Home the Bits* (Washington, D.C.: National Academy Press, 2002), pp. 144–46.

26. Scientists and telecommunication engineers have long been familiar with this landscape, but designers have only recently begun to give it the attention it deserves. See, for example, Anthony Dunne, *Hertzian Tales: Electronic*

Products, Aesthetic Experience, and Critical Design (London: Royal College of Art, 1999), and Anthony Dunne and Fiona Raby, *Design Noir: The Secret Life of Electronic Objects* (Basel: Birkhauser, 2001).

27.   There was a sudden fashion for warchalking in the summer of 2002, as 802.11 networks grew in popularity. It drew, rather self-consciously, upon the old hobo tradition of cryptically marking spots where hospitality might be expected, and maybe upon a central conceit of Thomas Pynchon's *The Crying of Lot 49*. The terminology derives from "wardialing" in the 1983 sci-fi movie *War Games*. See Jane Black, "A Wireless End Run Around ISPs," *BusinessWeek Online*, 3 July 2002, <www.businessweek.com> (accessed December 2002).

28.   For a discussion of this competition, and an analysis of related policy issues, see Spectrum Policy Task Force, *Report*, Federal Communications Commission, ET Docket No. 02-135, November 2002. The report urges reform of spectrum allocation policy to provide for more efficient use of spectrum.

29.   David Reed, "How Wireless Networks Scale: The Illusion of Spectrum Scarcity," *International Symposium on Advanced Radio Technology*, Boulder, 2 March 2002, <www.reed.com/dpr.html> (accessed December 2002).

30.   Leonard Kleinrock, "Breaking Loose," *Communications of the ACM* 44, no. 9 (September 2001): 41–45. See also Eric Knorr, "Mobile Web vs. Reality," *Technology Review* 104, no. 5 (June 2001): 56–63.

31.   Reliable wireless connection is one crucial technical component of nomadicity, but as Kleinrock points out, there are many others as well. For a survey of key technical issues, and some pioneering efforts to resolve them, see Guruduth Banavar and Abraham Bernstein, "Software Infrastructure and Design Challenges for Ubiquitous Computing Applications," *Communications of the ACM* 45, no. 12 (December 2002): 92–96.

32.   A pioneering example of spatially indefinite, self-configuring networking was Berkeley's MICA system for wireless sensor networks. See Mike Horton et al., "MICA: The Commercialization of Microsensor Motes," *Sensors Online*, April 2002, <www.sensorsmag.com> (accessed December 2002).

33.   Some pioneering analyses of these effects are provided in Barry Brown, Nicola Green, and Richard Harper, eds., *Wireless World: Social and Interactional Aspects of the Mobile Age* (London: Springer, 2002); J. Katz and M. Aakhus, *Perpetual Contact: Mobile Communications, Private Talk, and Public Performance* (Cambridge: Cambridge University Press, 2002); and Leysia Palen, "Mobile Telephony in a Connected Life," *Communications of the ACM* 45, no. 3 (March 2002): 78–82.

34.   There have been many accounts of this process. A particularly cogent, recent one is given by Jared Diamond in *Guns, Germs, and Steel: A Short History of Everybody for the Last 13,000 Years* (New York: Vintage, 1998). See, in particular, chapter 4, "Farmer Power."

35.   Jeremy Rifkin propounded a similar thesis, and maybe pushed it a bit too far, in *The Age of Access* (New York: Jeremy P. Tarcher, Putnam, 2000).

36. For a snapshot of emerging RFID-based commerce, circa 2002, see Amy Cortese, "Tollbooth Technology Meets the Checkout Lane," *New York Times*, 7 July 2002, section 3, p. 4.

37. For an early, classic analysis of the abstraction of money and its implications, see Georg Simmel, *The Philosophy of Money* (1907; London, Routledge, 1990).

38. Discussions of the cyborg subject and of cyborgs as social actors have appeared with increasing frequency in recent years. Gregory Bateson may have begun it with his speculations about an ecology of mind in the 1960s and early 1970s. One vigorous stream of theorizing, known as actor network theory, is rooted in the works of Michel Callon and Bruno Latour; see Bruno Latour, *We Have Never Been Modern* (Cambridge: Harvard University Press, 1993). Donna J. Haraway provided an influential feminist perspective in *Simians, Cyborgs, and Women: The Reinvention of Nature* (New York: Routledge, 1991) and emphasized the potentially disruptive and liberating qualities of the cyborg condition. Chris Hables Gray assembled key cyborg texts in *The Cyborg Handbook* (New York: Routledge, 1995). N. Katherine Hayles explored the relationship of subjectivity to the embodiment and disembodiment of information in *How We Became Posthuman: Virtual Bodies in Cybernetics, Literature, and Informatics* (Chicago: University of Chicago Press, 1999). And cyborg perspectives have begun to inform the literature of urban studies and planning; see Stephen Graham and Simon Marvin, *Splintering Urbanism: Networked Infrastructures, Technological Mobilities and the Urban Condition* (New York: Routledge, 2001).

39. Donna Haraway initiated a stream of critical discussion of the cyborg body, gender, and feminism in *Simians, Cyborgs, and Women*. On the construction of race in cyberspace, see Lisa Nakamura, *Cybertypes: Race, Ethnicity, and Identity on the Internet* (New York: Routledge, 2002).

40. Vernor Vinge anticipated the importance of this in his 1981 novella *True Names*, which begins:

> In the once-upon-a-time days of the First Age of Magic, the prudent sorceror regarded his own true name as his most valued possession but also the greatest threat to his continued good health, for—the stories go— once an enemy, even a weak unskilled enemy, learned the sorcerer's true name, then routine and widely known spells could destroy or enslave even the most powerful. As times passed, and we graduated to the Age of Reason and thence to the first and second industrial revolutions, such notions were discredited. Now it seems that the wheel has turned full circle (even if there never really was a First Age) and we are back to worrying about true names again.

Like Vinge's Mr. Slippery, Internet crackers and mischief-makers hide behind their online pseudonyms and worry that they will get busted if their true names are revealed. (For a reprint of *True Names*, together with commentaries on online identity and concealment, see James Frenkel ed., *True Names and the Opening of the Cyberspace Frontier* [New York: Tor, 2001]).

41. In his highly skeptical *On the Internet* (New York: Routledge, 2001), p. 11, Hubert L. Dreyfus has related this point specifically to the hyperlinked network structure of the World Wide Web. He observes: "Clearly the user of a hyper-connected library would no longer be a modern subject with fixed identity who desires a more complete and reliable model of the world, but rather a postmodern, protean being ready to be opened up to ever new horizons. Such a being is not interested in *collecting* what is *significant* but in *connecting* to *as wide a web of information as possible*."

42. Mark C. Taylor, *The Moment of Complexity: Emerging Network Culture* (Chicago: University of Chicago Press, 2002), p. 231. Somewhat similarly, Brian Massumi has suggested, in *Parables for the Virtual* (Durham: Duke University Press, 2002), p. 128: "The ex-human is now a node among nodes. Some nodes are still composed of organic body-matter, some are silicon-based, and others, like the ancestral robot arm, are alloy. The body-node sends, receives, and transduces in concert with every other node. The network is infinitely self-connectible, thus infinitely plastic. The shape and directions it takes are not centrally decided but emerge from the complex interplay of its operations."

43. In similar recognition of this point, there is now an emerging field of "cyborg anthropology"—the study of cultures in which the very definition of "man" is put into question by scientific and technological developments. See Joseph Dumit, Gary Lee Downey, and Sarah Williams, "Cyborg Anthropology," *Cultural Anthropology* 10, no. 2 (1995): 2–16; and Joseph Dumit and Gary Lee Downey, *Cyborgs and Citadels: Anthropological Interventions in Emerging Sciences and Technologies* (Santa Fe: School of American Research Press, 1997).

CHAPTER 4   DOWNSIZED DRY GOODS

1. Ryszard Kapuscinski, *The Shadow of the Sun* (New York: Knopf, 2001), pp. 229–30.

2. Decca sales literature made much of this in the postwar years:

> What did you do in the Great War—"Decca"?
> I was "Mirth-Maker-in-Chief to His Majesty's Forces"; my role being to give our Soldiers and our Sailors music wherever they should be. In that capacity I saw service on every Front—France, Belgium, Egypt, Palestine, Italy and the Dardenelles; Right in the Front Line and away back in Camps and Hospitals. All told, there were 100,000 "Deccas" on Active Service from Start to Finish of the War.

Quoted in *Jones Telecommunications and Multimedia Encyclopedia*, <www.digitalcentury.com> (accessed December 2002).

3. Since 1992, the Semiconductor Industry Association (SIA) has been monitoring the shrinkage of integrated circuit elements and publishing

predictions. The SIA's *International Technology Roadmap for Semiconductors 2002* <http://public.itrs.net/> (accessed March 2003) suggested that the limits were only a few years away.

4. They are also called microsystems and micromechatronic systems. For a snapshot of MEMS technology in the early 2000s, and some projections of future directions, see Committee on Implications of Emerging Micro- and Nanotechnologies, *Implications of Emerging Micro- and Nanotechnologies* (Washington, D.C.: National Academies Press, 2002).

5. Richard Feynman, "There's Plenty of Room at the Bottom," *Engineering and Science* 23 (February 1960): 22–36.

6. K. Eric Drexler, *Engines of Creation* (Garden City, N.Y.: Anchor, 1987). See also K. Eric Drexler, *Nanosystems: Molecular Machinery, Manufacturing, and Computation* (New York: Wiley Interscience, 1992).

7. Committee for the Review of the National Nanotechnology Initiative, *Small Wonders, Endless Frontiers* (Washington, D.C.: National Academy Press, 2002).

8. For example, *Scientific American* provided a comprehensive, critical overview in a September 2001 special issue entitled "Nanotech: The Science of Small Gets Down to Business," vol. 285, no. 3.

9. Michael Crichton, *Prey* (New York: HarperCollins, 2002).

10. L. A. Bumm, J. J. Arnold, M. T. Cygan, T. D. Dunbar, T. P. Burgin, L. Jones II, D. L. Allara, J. M. Tour, and P. S. Weiss, "Are Single Molecular Wires Conducting?," *Science* 271 (22 March 1996): 1705–07.

11. Mark A. Kastner, "Artificial Atoms," *Physics Today* 46 (January 1993): 24–31. See also Richard Turton, *The Quantum Dot: A Journey into the Future of Microelectronics* (Oxford: Oxford University Press, 1995).

12. David Rotman, "The Nanotube Computer," *Technology Review* 105, no. 2 (March 2002): 37–45.

13. Harold Abelson, Don Allen, Daniel Coore, Chris Hanson, George Homsy, Thomas F. Knight, Radhika Nagpal, Erik Rauch, Gerald Jay Sussman, and Ron Weiss, "Amorphous Computing," *Communications of the ACM* 43, no. 5 (May 2000): 74–82.

14. Kimberly Hamad-Schifferli, John J. Schwartz, Aaron T. Santos, Shuguang Zhang, and Joseph M. Jacobson, "Remote Electronic Control of DNA Hybridization through Inductive Coupling to an Attached Metal Nano-crystal Antenna," *Nature* 415 (10 January 2002): 152–55.

15. Lawrence K. Altman, "Self-Contained Mechanical Heart Throbs for First Time in a Human," *New York Times*, 4 July 2001, pp. A1, A10.

16. Feynman, "There's Plenty of Room at the Bottom," pp. 22–36. Medical nano-technology has since become an active research field; see Robert A. Freitas Jr., *Nanomedicine, Volume 1: Basic Capabilities* (Georgetown: Landes Bioscience, 1999), and <www.nanomedicine.com> (accessed December 2002).

17. Ray Kurzweil, "The Singularity: A Talk with Ray Kurzweil," *Edge* 99 (25 March 2002), <www.edge.org/documents/archive/edge99.html> (accessed December 2002).

18. Their modern descendants, server farms, continue this tradition—but in modular rack rather than cabinet format.

19. Various subgenres emerged during the 1980s and 1990s—the tower, the cube, the pizza box, and so on; designers mostly focused on providing a stylish outer surface. In some models (such as the first Macintosh) the processor and monitor were integrated into a single box, while in others they became separate elements. As processor boxes shrank, and as thin plasma screens began to replace bulky CRTs, designers had greater freedom to shape these elements; Apple, in particular, took adventurous advantage of this possibility.

20. Japan led the way. J-Phone introduced Sha-mail (email with photographs) handsets in November 2000. They were soon followed by NTT DoCoMo and others. KDDI and Casio introduced a GPS-equipped model in May 2002. An increasing percentage of Japanese cellphones began to incorporate cameras, and there was speculation that sales of these hybrids would eventually outstrip those of conventional digital cameras.

21. In 2000/2001, the Kyocera Smartphone, the Samsung 1300, and the Handspring Treo were among the first successful products of this type.

22. For a detailed analysis of the "spaces on the human body where solid and flexible forms can rest," with particular reference to the design of wearable electronic devices, see Francine Gemperle, Chris Kasabach, John Stivoric, Malcolm Bauer, and Richard Martin, "Design for Wearability," *Proceedings of the Second International Symposium on Wearable Computers* (Los Alamitos, Calif.: IEEE Computer Society Press, 1998), pp. 116–22.

23. MIT news release, 14 March 2002, <http://web.mit.edu/newsoffice/nr/2002/isn.html> (accessed December 2002).

24. Catherine Dike, *Cane Curiosa: From Gun to Gadget* (Paris: Les Editions de l'Amateur, 1983).

25. Applied Digital Solutions began to market the implantable VeriChip, which held an identification number that could be read by a scanner, in 2002. University of Reading computer scientist Kevin Warwick gained considerable notoriety (if not research results) by having himself implanted with microelectronics in the early 2000s; see Kevin Warwick, *I, Cyborg* (London: Century, 2002).

26. MIT news release, ibid. The appropriation of images and scenarios from superhero comics was direct; the *Boston Globe* (28 August 2002, pp. A1, A6) gleefully reported that an image on the Institute for Soldier Nanotechnologies's Web site had been redrawn from the Radix comic book character Valerie Fiores—an armor-clad security officer with attitude.

27. For a snapshot of smart fabrics research and development efforts in the early 2000s, see Lori Valigra, "Fabricating the Future," *Christian Science Monitor*, 29 August 2002, <www.csmonitor.com> (accessed December 2002).

28. Duncan Graham-Rowe, "Remote Heartbeat Monitor Unveiled," NewScientist.com, 28 January 2002. (*Measurement Science and Technology*, vol. 13, p. 163.)

29. For a comprehensive introduction to the functions of clothing systems, see Susan M. Watkins, *Clothing: The Portable Environment*, 2d ed. (Ames: Iowa State University Press, 1995).

CHAPTER 5 SHEDDING ATOMS

1. As information economists are careful to point out, however, nonrival assets can certainly have market value. Electronically distributed stock prices, for example, can be reproduced at negligible cost. They have high value when they are up to the second and exclusive but rapidly decline in value with time and with wider distribution. For a detailed discussion of the differences between rival and nonrival assets, and the implications of these, see Lawrence Lessig, *The Future of Ideas: The Fate of the Commons in a Connected World* (New York: Random House, 2001).

2. Fred Lerner, *The Story of Libraries: From the Invention of Writing to the Computer Age* (New York: Continuum, 1998); Lionel Casson, *Libraries in the Ancient World* (New Haven: Yale University Press, 2001).

3. Henry Petroski, *The Book on the Bookshelf* (New York: Knopf, 1999).

4. Predictably, this eventually produced a bibliophile backlash. See Nicholson Baker, *Double Fold: Libraries and the Assault on Paper* (New York: Random House, 2001).

5. Vannevar Bush, "As We May Think," *Atlantic Monthly* 176, no. 1 (July 1945): 101–8.

6. <www.tlg.uci.edu> (accessed December 2002).

7. If you want to get picky about the physics, we can say that the corpus of classical literature is now embodied electromagnetically, and yes, electrons do have mass. But that is irrelevant at the level of everyday experience. My briefcase quickly gets weighed down if I load volumes of the Loeb Classical Library into it, but my laptop does not get any heavier if I download the *TLG* onto its hard drive.

8. <arXiv.org> (accessed December 2002). See also James Glanz, "The World of Science Becomes a Global Village," *New York Times*, 1 May 2001, D1–D2.

9. Paul Ginsparg, "Electronic Clones vs. the Global Research Archive," <arXiv.org/blurb/pg00bmc.html> (accessed December 2002). Ginsparg and the archive later moved to Cornell. For a brief history of arXiv, see Gary Stix, "Wired Superstrings," *Scientific American* 288, no. 5 (May 2003): 38–39.

10. <CogNet.mit.edu> (accessed December 2002). See also Marney Smyth, "The Community *Is* the Content," *Publishing Research Quarterly* 17, no. 1 (Spring 2001): 3–14.

11. <www.ArchNet.org> (accessed December 2002).

12. See, for example, MIT's DSpace, <web.mit.edu/dspace> (accessed December 2002).

13. National Research Council, *LC21: A Digital Strategy for the Library of Congress* (Washington, D.C.: National Academy Press, 2000), p. 3. On prospects

for a global digital library, see Christine Borgman, *From Gutenberg to the Global Information Infrastructure: Access to Information in the Networked World* (Cambridge: MIT Press, 2000).

14. Due to the topological properties of the Web and the inherent limitations of Web search engines, we know that this can only be a fraction of the total Web. (See Steve Lawrence and C. Lee Giles, "Searching the World Wide Web," *Science* 280 [1998]: 98–100.) The Internet Archive is even larger; last time I looked it claimed 100 terabytes.

15. Reme Ahmad, "Malaysia Bans Use of SMS for Divorces," *Straits Times*, 13 July 2001, p. A8.

16. See for example Audit Commission, *A Stitch in Time: Facing the Challenge of the Year 2000 Date Change* (London: Audit Commission, 1998), <www.y2kbug.org.uk> (accessed December 2002).

17. I noted that "code is the law" in *City of Bits* (Cambridge: MIT Press, 1995), p. 111. Lawrence Lessig subsequently provided a much more extensive and detailed treatment of the idea in *Code and Other Laws of Cyberspace* (New York: Basic Books, 1998). Then the Y2K scare motivated a large-scale effort to locate code and evaluate the consequences of its failure, so providing a useful benchmark of the power that code had attained by the end of 1999. If you want, you can give this point the full Foucault treatment, but I will leave that as an exercise for the reader.

18. Philip Steadman, *Vermeer's Camera: Uncovering the Truth behind the Masterpieces* (Oxford: Oxford University Press, 2001).

19. Rum—being one of the few things of value around—served as currency in the early days of the colony of Sydney. It functioned well enough as a means of exchange, a unit of account (when you put it in containers), and a store of value. Instead of running up bar tabs, you could directly drink away your liquid assets. And here, the forger's option was to water the currency down. See Ross Fitzgerald and Mark Hearn, *Bligh, Macarthur, and the Rum Rebellion* (Kenthurst: Kangaroo Press, 1988).

20. David S. Evans and Richard Schmalensee, *Paying with Plastic: The Digital Revolution in Buying and Borrowing* (Cambridge: MIT Press, 1999).

21. Evan I. Schwartz, "How You'll Pay," *Technology Review* 105, no. 10 (December 2002/January 2003): 50–57.

22. For a useful survey of telecommunications in banking and financial markets, circa 2000, see John F. Langdale, "Telecommunications and 24-Hour Trading in the International Securities Industry," in Mark I. Wilson and Kenneth E. Corey, eds., *Information Tectonics: Space, Place and Technology in an Electronic Age* (Chichester: Wiley, 2000), pp. 89–99. For a detailed snapshot of the emerging technology of digital money, as it appeared in the early days of Internet commerce, see Daniel C. Lynch and Leslie Lundquist, *Digital Money: The New Era of Internet Commerce* (New York: John Wiley, 1996). For longer-range, more speculative views of the digital future of money, see Kevin Kelly,

"E-Money," chapter 12 in *Out of Control* (Cambridge: Perseus Books, 1994), and Neil Gershenfeld, "Smart Money," in *When Things Start to Think* (New York: Henry Holt, 1999), pp. 78–91.

23. Nelson Goodman, *Languages of Art*, 2d ed. (Indianapolis: Hackett, 1976), p. 113.

24. Recognizing that many users would download files over slow lines, the designers of MP3 chose a compressed format. Digital compression is one of the newer forms of lightening in order to speed movement. Files are digitally processed to reduce their size for transmission, then reprocessed at the receiving end to decompress them. This entails some loss of quality, but it is within acceptable limits for most purposes.

25. For a general introduction to peer-to-peer systems and associated issues, see Andy Oram, ed., *Peer-to-Peer: Harnessing the Benefits of a Disruptive Technology* (Sebastopol, Calif.: O'Reilly, 2001).

26. By July 2001 the music industry had succeeded in neutralizing Napster. But by then a new generation of even more effective online exchange systems had appeared. (See Matt Richtel, "With Napster Down, Its Audience Fans Out," *New York Times*, 20 July 2001, pp. A1, C2.) Furthermore, the idea was beginning to extend to movie and video distribution; by March 2002 MusicCity. com's Morpheus system, KaZaA, and others were supporting online video file swapping, and the Motion Picture Association of America was filing suit in Los Angeles federal district court—claiming that these services constituted a "21st century piratical bazaar where the unlawful exchange of protected materials takes place across the vast expanses of the Internet."

CHAPTER 6 DIGITAL DOUBLIN'

1. These sites are defined by their inhabitants and attractions and, incidentally (as all Joyceans know), map onto the settings of Homer's *Odyssey*. The action unfolds successively through the tower, school, strand, house, bath, graveyard, newspaper office, restaurant, library, streets, concert room, tavern, rocks, hospital, brothel, shelter, house, and bed. There have been numerous explications of the mapping of these settings onto actual Dublin sites; see, for example, Cyril Pearl, *Dublin in Bloomtime: The City James Joyce Knew* (New York: Viking, 1969); Robert Nicholson, *The Ulysses Guide: Tours through Joyce's Dublin* (London: Routledge, 1989); and Rosanna Negrotti, *Joyce's Dublin: An Illustrated Commentary* (London: Caxton Editions, 2000).

2. "What facilities of transit were desirable?" Bloom asked himself. The reply:

> When citybound frequent connection by train or tram from their respective intermediate station or terminal. When countrybound velocipedes, a chainless freewheel roadster cycle with side basketcar attached, or

draught conveyance, a donkey with wicker trap or smart phaeton with good working solidungular cob (roan gelding, 14h).

James Joyce *Ulysses* (New York: Everyman's Library, 1997), pp. 969–70.

3.   Ibid, p. 974.

4.   Kevin Livingston, "Communications," in Graeme Davison, John Hirst, and Stuart Macintyre, eds., *The Oxford Companion to Australian History* (Melbourne: Oxford University Press, 1998), p. 143. See also Kevin Livingston, *The Wired Nation Continent: The Communication Revolution and Federating Australia* (Melbourne: Oxford University Press, 1997), for an account of "technological nationalism" in Australia.

5.   Michael Wines, "Wired Radio Offers Fraying Link to Russian Past," *New York Times*, 18 October 2001, p. A4.

6.   Russ Hodges, Giants play-by-play broadcaster, New York, 3 October 1951.

7.   "Radio Listeners in Panic, Taking War Drama as Fact," *New York Times*, 31 October 1938, p. 1. Subsequent commentators have pointed out that the newspapers, conscious of competition from the new medium of radio, probably exaggerated the panic a bit to discredit the rival medium. Welles himself remarked that he had hesitated about presenting it because "it was our thought that perhaps people might be bored or annoyed at hearing a tale so improbable."

8.   Hugh Kenner, *Dublin's Joyce* (New York: Columbia University Press, 1987), pp. 165–68.

9.   The film actually presents a composite of Moscow, Kiev, Odessa, and elsewhere.

10.   Kenner, *Dublin's Joyce*, pp. 167–68.

11.   Steve Mann, "Wearable Computing: A First Step Toward Personal Imaging," *IEEE Computer* 30, no. 2 (February 1997): 25–32; and Steve Mann, *Intelligent Image Processing* (New York: Wiley, 2001).

12.   See Alexi Worth, "Wolfgang Staehle, Untitled, 2001," *Artforum* (November 2001), p. 129, and Nancy Princenthal, "Wolfgang Staehle at Postmasters," *Art in America* (November 2001), p. 142.

13.   David Gelernter popularized a version of this idea, before the explosion of the World Wide Web and the proliferation of wireless devices, in *Mirror Worlds, or the Day Software Puts the Universe in a Shoebox* (New York: Oxford University Press, 1991). For more recent commentary see Philip E. Agre, "Beyond the Mirror World: Privacy and the Representational Practices of Computing," in Philip E. Agre and Marc Rotenberg, eds., *Technology and Privacy: The New Landscape* (Cambridge: MIT Press, 1998).

CHAPTER 7   ELECTRONIC MNEMOTECHNICS

1.   Today, of course, electronic help is available: Islamsoft's Prayer Time 4.0 calculates prayer times for specified locations and provides notification, World

Qiblah indicates the direction of Mecca, and USA Masjid Locator finds nearby mosques. See Toby Lester, "Guiding Light," *Atlantic Unbound*, 13 January 1999, <www.theatlantic.com/unbound/citation/wc990113.htm> (accessed December 2002).

2. Location awareness is an aspect of context awareness—the ability to determine and make use of contextual information such as location, time and date, temperature, sound level, surrounding objects and people, and so on. Context-aware computing has been an active research field since the early 1990s. For an introduction and survey, see Guanling Chen and David Kotz, "A Survey of Context-Aware Mobile Computing Research," Dartmouth Computer Science Technical Report TR2000–381, 2000. On the interrelationships of mobility, wearability, and context-awareness, see Daniel P. Siewiorek, "New Frontiers in Application Design," *Communications of the ACM* 45, no. 12 (December 2002): 79–82.

3. Pioneering demonstrations of location-aware computing and communication were the Olivetti Research Active Badge system (Roy Want, Andy Hopper, Veronica Falcao, and Jonathan Gibbons, "The Active Badge Location System," *ACM Transactions on Information Systems* 10, no. 1 [January 1992]: 91–102) and the Xerox PARC PARCTAB system (Roy Want, Bill N. Schilit, Norman I. Adams, Rich Gold, Karin Petersen, David Goldberg, John R. Ellis, and Mark Weiser, "An Overview of the PARCTAB Ubiquitous Computing Experiment," *IEEE Personal Communications* 2, no. 6 [December 1995]: 28–43.) The seminal discussion of the uses of location-aware computing is Mark Weiser, "The Computer for the 21st Century," *Scientific American* 265, no. 3 (September 1991): 94–104.

4. The Swedish National Road Administration has pioneered the use of location-aware, customized warning systems on automobile dashboards. See Julie Claire Diop, "Sensing Speed Limits," *Technology Review* 105, no. 10 (December 2002/January 2003): 29.

5. This has, of course, raised considerable civil liberties concerns. See, for example, Stuart Millar and Paul Kelso, "Liberties Fear over Mobile Phone Details," *Guardian*, 27 October 2001; and Elizabeth Douglass, "Cell Phones Set to Track Call Locales," *Los Angeles Times*, 18 October 2001.

6. <http://www.uwb.org> (accessed December 2002).

7. For further details, see <http://gps.faa.gov> (accessed December 2002).

8. For even greater accuracy, quantum-enhanced procedures can potentially be used. See Vittorio Giovannetti, Seth Lloyd, and Lorenzo Maccone, "Quantum-Enhanced Positioning and Clock Synchronization," *Nature* 412 (2001): 417–19.

9. In the mid-1990s, the U.S. Federal Communications Commission passed a plan to provide 911 operators with more precise cellphone call locations. Implementation turned out, however, to be a slow and difficult matter. In Switzerland, the FriendZone system pioneered location-based cellphone service in the early 2000s.

10. Nissanka B. Priyantha, Anit Chakraborty, and Hari Balakrishnan, "The Cricket Location-Support System," *6th ACM Conference on Mobile Computing and Networking (ACM MOBICOM)*, Boston, August 2000.

11. Paul Virilio, *Strategy of Deception* (London: Verso, 2000), p. 33.

12. See, for example, Adam Clymer, "Tracking Bay Area Traffic Creates Concern for Privacy," *New York Times*, 26 August 2002, p. A10.

13. Sanjay Sarma, David Brock, and David Engels, "Radio Frequency Identification and the Electronic Product Code," *IEEE Micro* 21, no. 6 (November/December 2001): 50–54. By 2003, RFID technology was taking off; Gillette, for example, was embedding RFID tags in razor packaging, and ordering hundreds of millions of tags.

14. Paul Rogers, "Your Future Car May Be a Spy: Clean-Air Proposal Raises Privacy Concerns," *San Jose Mercury News*, 14 June 1996.

15. See, for example, the work of the Cambridge/MIT Auto-ID Center, <www.autoid.org> (accessed December 2002). Note, in particular, Duncan McFarlane, "Auto-ID Based Control," white paper, 1 February 2002.

16. In the aftermath of the World Trade Center attack in September 2001, FBI efforts to track the preparations of the terrorists vividly revealed the extent of our growing electronic visibility. Investigators put together a detailed picture from records of cellphone calls, credit card transactions, automated teller machine withdrawals, and communications through email and Internet chat rooms. See Kate Zernike and Don Van Natta Jr., "Hijackers' Meticulous Strategy of Brains, Muscle and Practice," *New York Times*, 4 November 2001, pp. A1 and B6.

17. For example, the sentence "The Empire State Building is tall," unambiguously refers to an actual object on Fifth Avenue and informs us about its height. And it's true: you can look it up. Things get trickier with "The prettiest building on Fifth Avenue is tall," "The duck-shaped, polka-dotted building on Fifth Avenue is tall," and "New York is Gotham."

18. Robert Venturi, Denise Scott Brown, and Steven Izenour, *Learning from Las Vegas* (Cambridge: MIT Press, 1972).

19. Robert Venturi and Denise Scott Brown, "Las Vegas after Its Classic Age," in Robert Venturi, *Iconography and Electronics upon a Generic Architecture* (Cambridge: MIT Press, 1996), p. 126.

20. On the nonobjectivity of maps, see J. B. Harley, *The New Nature of Maps: Essays in the History of Cartography* (Baltimore: Johns Hopkins University Press, 2001).

21. Provincetown's traditional way of dealing with this issue is the display, or nondisplay, of rainbow flags outside establishments—much like the hobo tradition of cryptic chalked marks. But such visible declaration causes certain tensions; there is an argument for the discretion of wireless.

22. It is bad enough that "The Morning Star is the Evening Star." But you are in even bigger trouble if you cannot figure out which point of light in the heavens is supposed to be the Morning Star.

23. For an introduction to GIS technology and its applications, see Nicholas Chrisman, *Exploring Geographic Information Systems*, 2d ed. (New York: Wiley, 2001).

24. Street patterns change over time, so automobile navigation maps require updating (see Ivan Berger, "Does Broadway Still Meet 42nd St.?" *New York Times*, 3 January 2003, p. D8). Early navigation systems stored the maps on CDs, which could be replaced at intervals with updated versions. More advanced systems might provide for wireless downloads, as needed, of current maps.

25. Metadata is content about content. Typically it associates with a document such information as a heading, author's name, date and place of publication, classification, and keywords. It can also provide spatial coordinates, such as the originator's street address, or the latitude and longitude of a building that is described. For an introduction to metadata and its uses, see Christine L. Borgman, "Access to Information," chapter 3 in *From Gutenberg to the Global Information Infrastructure: Access to Information in the Networked World* (Cambridge: MIT Press, 2000), pp. 53–80.

26. There have been several implementations of location-aware tourist and exhibition guide systems. See, for example, Gregory D. Abowd, Christopher G. Atkeson, Jason Hong, Sue Long, Rob Kooper, and Mike Pinkerton, "Cyberguide: A Mobile Context-Aware Tour Guide," *Wireless Networks* 3, no. 5 (October 1997): 421–33; Nigel Davis, Keith Cheverst, Keith Mitchell, and Adrian Friday, "Disseminating Tourist Information in the GUIDE System," *Proceedings of the Second IEEE Workshop on Mobile Computing Systems and Applications* (New Orleans: IEEE Computer Society Press, 1999); Benjamin B. Bederson, "Audio Augmented Reality: A Prototype Automated Tour Guide," *Proceedings of Conference on Human Factors and Computing Systems, CHI '95* (Denver: ACM Press, May 1995), pp. 210–211; and Reinhard Oppermann and Marcus Specht, "A Context-Sensitive Nomadic Exhibition Guide," *Proceedings of Second International Symposium on Handheld and Ubiquitous Computing, HUC 2000* (Bristol, England: Springer-Verlag, 2000), pp. 127–42.

27. Alorie Gilbert, "Smart Carts on a Roll at Safeway," CNET News.com, 28 October 2002, <http://news.com.com/2100-1017-963526.html?tag= fd_top> (accessed December 2002). On the technology of location-aware shopping systems, see Abhaya Asthana, Mark Cravatts, and Paul Kryzanowski, "An Indoor Wireless System for Personalized Shopping Assistance," *Proceedings of IEEE Workshop on Mobile Computing Systems and Applications* (Santa Cruz: IEEE Computer Society Press, 1994), pp. 69–74.

28. See, however, Michael Deer, Nigel Thrift, and Derek Gregory, eds., *Ground Truth: The Social Implications of Geographic Information Systems* (New York: Guilford Press, 1995); Jon Goss, "We Know Who You Are and We Know Where You Live: The Instrumental Rationality of Geodemographic Systems," *Economic Geography* 71 (1995): pp. 171–98; and Mark Monmonier, *Spying with Maps* (Chicago: University of Chicago Press, 2002).

29. Wearable retinal scanning displays have, for example, been developed and marketed by Microvision. See <www.mvis.com> (accessed December 2002). On the development and importance of miniaturized see-through displays, see Erik Sherman, "Little Big Screen," *Technology Review* 104, no. 5 (June 2001): 64–69.

30. Ronald T. Azuma, Yohan Baillot, Reinhold Behringer, Steven K. Feiner, Simon Julier, and Blair MacIntyre, "Recent Advances in Augmented Reality," *IEEE Computer Graphics and Applications* 21, no. 6 (November/December 2001): 34–47, and Steven K. Feiner, "Augmented Reality: A New Way of Seeing," *Scientific American* 286, no. 4 (April 2002): 48–55.

31. Jean Baudrillard, *Simulations* (New York: Semiotext(e), 1983).

32. Cicero, *De oratore*, II, lxxxvi. Frances A. Yates tells the tale to introduce her classic study, *The Art of Memory* (Chicago: University of Chicago Press, 1966), pp. 1–2.

33. Yates, *Art of Memory*, p. 3.

34. Early examples of systems and proposals of this sort include GeoNotes, which advertises itself as "digital graffiti in public places," <http://geonotes.sics.se/> (accessed December 2002); WorldBoard (J. C. Spohrer, "Information in Places," *IBM Systems Journal* 38, no. 4 (1999), <www.research.ibm.com/journal/sj/384/spohrer.html> [accessed December 2002]); Graffiti, described as "a simple application which allows users to leave notes for people at various locations on the Cornell campus," <www.cs.cornell.edu/boom/2001sp/kubo/egraffiti.html> (accessed December 2002); CampusAware, <http://testing.hci.cornell.edu/context/campus_aware/> (accessed December 2002); Bits on Location, <www.datenamort.de> (accessed December 2002); MemoClip (Michael Beigl, "MemoClip: A Location Based Remembrance Appliance," <www.teco.edu/~michael/publication/memoclip.pdf> [accessed December 2002]); and 34 North 118 West, <http://34n118w.net/htmldir/descriptn.html> (accessed December 2002).

CHAPTER 8 FOOTLOOSE FABRICATION

1. Thoreau observed the ice harvest on Walden Pond and, characteristically, was not too happy about this intrusion of industry. The story of the New England ice industry is entertainingly told in Gavin Weightman, *The Frozen Water Trade: How Ice from New England Lakes Kept the World Cool* (London: HarperCollins, 2001).

2. Amy Harmon, "Music Industry in Global Fight on Web Copies," *New York Times*, 7 October 2002, pp. A1, A6; and Ariana Eunjung Cha, "File Swapper Eluding Pursuers," *Washington Post*, 21 December 2002, p. A1.

3. William J. Mitchell, "Roll Over Euclid: How Frank Gehry Designs and Builds," in J. Fiona Ragheb, ed., *Frank Gehry, Architect* (New York: Abrams, 2001), pp. 352–64.

4. For an early discussion of this development, from a publisher's viewpoint, see Jason Epstein, "Reading: The Digital Future," *New York Review of Books* 48, no. 11 (July 5, 2001): 46–48. On the future of books, publishers, booksellers, and libraries more generally, see Clifford Lynch, "The Battle to Define the Future of the Book in the Digital World," *First Monday* 6, no. 6 (June 2001), <www.firstmonday.org/issues/issue6_6/lynch/index.html> (accessed December 2002).

5. Brent A. Ridley, Babak Nivi, and Joseph M. Jacobson, "All-Inorganic Field Effect Transistors Fabricated by Printing," *Science* 286 (22 October 1999): 746–49; Stephen Mihm, "Print Your Next PC," *Technology Review* 103, no. 6 (November/December 2000): 66–70; and Joseph Jacobson, "The Desktop Fab," *Communications of the ACM* 44, no. 3 (March 2001): 41–42.

6. The possibility of designing your own shoes on the Web was pioneered by cmax.com. Other early customization sites offered computers (dell.com), automobiles (mini.com), kitchens (merillat.com), watches (sega.com), and homes (lindal.com).

7. Walter Benjamin, "The Work of Art in the Age of Mechanical Reproduction," in *Illuminations*, trans. Harry Zohn (New York: Schocken, 1969), pp. 217–51.

8. National Research Council, *The Digital Dilemma: Intellectual Property in the Information Age* (Washington, D.C.: National Academy Press, 2000).

9. This practice was anticipated, with earlier and less convenient technology, in the production of compilation tapes.

10. For a more detailed discussion of this point, see William J. Mitchell, *The Reconfigured Eye: Visual Truth in the Post-Photographic Era* (Cambridge: MIT Press, 1992).

11. For a discussion of the roles of print, see Mario Carpo, *Architecture in the Age of Printing* (Cambridge: MIT Press, 2001).

12. For some examples, see William J. Mitchell, "Dream Homes," *New Scientist* 174, no. 2347 (15 June, 2002): 38–42.

13. Eric Raymond, *The Cathedral and the Bazaar: Musings on Linux and Open Source by an Accidental Revolutionary* (Sebastopol, Calif.: O'Reilly, 1999). Pekka Himanen, *The Hacker Ethic and the Spirit of the Information Age* (New York: Random House, 2001).

14. For a convenient introduction, with practical examples from a wide variety of fields, see Peter J. Bentley and David W. Corne, *Creative Evolutionary Systems* (San Francisco: Morgan Kaufmann, 2002).

15. In the United States, the most egregious outcome of these efforts has been the Digital Millennium Copyright Act.

16. Neal Stephenson, *The Diamond Age* (New York: Bantam, 1995).

1.   The image is, of course, anachronistic. It shows a fifteenth-century scholar rather than a fourth-century one. But this does not affect my point.

2.   Jerome's cube was illustrated by many medieval and Renaissance artists, including Antonello and Dürer. Dilbert's was chronicled by Scot Adams in a popular syndicated strip.

3.   Early descriptions of the process of digital fragmentation and recombination, as it had begun to unfold, were provided in William J. Mitchell, *City of Bits: Space, Place, and the Infobahn* (Cambridge: MIT Press, 1994), and William J. Mitchell, *E-topia: Urban Life, Jim—But Not As We Know It* (Cambridge: MIT Press, 1999).

4.   For a discussion of this issue, and the unintended consequences of the ban, see Jennifer 8. Lee, "Students with Gadgets a Hang-up for Teachers," *International Herald Tribune*, 17–18 August 2002, pp. 1, 5.

5.   For an early snapshot of this transformation in progress, see Josh McHugh, "Unplugged U," *Wired*, October 2002, <www.wired.com/wired/archive/10.10/dartmouth.html> (accessed December 2002).

6.   See, for example, Dadong Wan, "Magic Medicine Cabinet: A Situated Portal for Healthcare," *Proceedings of International Symposium on Handheld and Ubiquitous Computing* (Karlsruhe, September 1999).

7.   Andrew Fano and Anatole Gershman, "The Future of Business Services in the Age of Ubiquitous Computing," *Communications of the ACM* 45, no. 12 (December 2002): 83–87.

8.   For an introduction to telecommuting, live/work, and the associated literature, see William J. Mitchell, "Homes and Neighborhoods," chapter 5 in *E-topia*. A more recent and detailed overview is provided by Penny Gurstein, *Wired to the World, Chained to the Home: Telework in Daily Life* (Vancouver: University of British Columbia Press, 2001).

9.   For a useful typology of live/work space configurations, see Gurstein, *Wired to the World*, pp. 138–45.

10.   <www.parcbit.es> (accessed December 2002).

11.   Ian Wylie, "La Dolce Vita, Internet Style," *Fast Company*, 20 August 2002, p. 76. See also <www.colletta.it> (accessed December 2002).

12.   For an analysis of the potential advantages and disadvantages of mobilizing information work, see Gordon B. Davis, "Anytime/Anyplace Computing and the Future of Knowledge Work," *Communications of the ACM* 45, no. 12 (December 2002): 67–73.

13.   <www.remotelounge.com> (accessed December 2002).

14.   On the complex set of issues related to domestic connection, see National Research Council, *Broadband: Bringing Home the Bits* (Washington, D.C.: National Academy Press, 2002).

15.   For a lively discussion of Seoul's PC *baang* culture, in the early 2000s, see J. C. Hertz, "The Bandwidth Capital of the World," *Wired* 10, no. 8

(August 2002), <www.wired.com/wired/archive/10.08/korea.html> (accessed December 2002). See also Howard W. French, "Korea's Real Rage for Virtual Games," *New York Times*, 9 October 2002, p. A8.

16. This creates a demand for software to help arrange ad hoc meetings among mobile participants and to perform meeting follow-up functions. See, for example, Mikael Wiberg, "RoamWare: An Integrated Architecture for Seamless Interaction in between Mobile Meetings," *ACM Group '01* (Boulder, September 2001).

CHAPTER 10 AGAINST PROGRAM

1. Michel de Certeau, *The Practice of Everyday Life* (Berkeley: University of California Press, 1984).

2. Ibid., p. xxiv.

3. Walter Benjamin, "On Some Motifs in Baudelaire," in *Illuminations*, trans. Harry Zohn (New York: Schocken, 1969), pp. 155–200. See also Chris Jenks, "Watching Your Step: The History and Practice of the *Flâneur*," in Chris Jenks, ed., *Visual Culture* (London: Routledge, 1995), pp. 142–60.

4. This idea was given architectural expression in the "New Babylon" project by the Dutch artist Constant Nieuwenhuys, 1956–74. See Catherine de Zegher and Mark Wigley, eds., *The Activist Drawing: Retracing Situationist Architectures from Constant's New Babylon to Beyond* (Cambridge: MIT Press, 2001).

5. Gilles Deleuze and Félix Guattari, "Treatise on Nomadology:—The War Machine," in *A Thousand Plateaus*, trans. Brian Massumi (Minneapolis: University of Minnesota Press, 1987), pp. 351–423.

6. See, for example, Critical Art Ensemble, "Nomadic Power and Cultural Resistance," in *The Electronic Disturbance* (Brooklyn, N.Y.: Autonomedia, 1994), pp. 11–34.

7. One might extend Canetti's well known analysis to the digital wireless era. See Elias Canetti, *Crowds and Power* (New York: Seabury Press, 1982).

8. John Summerson, "The Case for a Theory of Modern Architecture," *Royal Institute of British Architects Journal* 64 (1957): 307–10, reprinted in John Summerson, *The Unromantic Castle and Other Essays* (London: Thames and Hudson, 1990), pp. 257–66. Summerson's definition runs as follows: "A program is a description of the spatial dimensions, spatial relationships and other physical conditions required for the convenient performance of specific functions . . . It is difficult to imagine any program in which there is not some rhythmically repetitive pattern—whether it is a manufacturing process, the curriculum of a school, the domestic routine of a house, or simply the sense of repeated movement in a circulation pattern."

9. Michael Batty, "Editorial: Thinking about Cities as Spatial Events," *Environment and Planning B*, 29 (2002): 1–2.

10.   In the period of transition from physical security of paper to electronic security of online files, the question of when and where you could download, and where you could keep electronic copies, suddenly became critical in high security settings. The cases of Wen Ho Lee and John Deutch turned on this point.

11.   For an overview of the technical issues involved in this see "Self-configuration and Adaptive Coordination," in National Research Council, *Embedded Everywhere: A Research Agenda for Networked Systems of Embedded Computers* (Washington, D.C.: National Academy Press, 2001), pp. 76–118.

12.   The nineteenth-century bourgeois were the inverse of the twenty-first century's emerging electronomads. In *History of Bourgeois Perception* (Chicago: University of Chicago Press, 1982), p. 71, Donald M. Lowe noted that "the bourgeoisie had a compulsion to fill up the visible space of the home with excessive furniture and intricate decoration. They cluttered every room in the house with objects. The eye seemed to abhor any visible, empty space."

13.   Henri Lefebvre, *Writings on Cities*, ed. Eleonore Kofman and Elizabeth Lebas (London: Blackwell, 1996), p. 195. See also Henri Lefebvre, "The Monument" and "The Space of Architects," in *The Production of Space*, trans. Donald Nicholson-Smith (London: Blackwell, 1991).

14.   Reyner Banham, Paul Barker, Peter Hall, and Cedric Price, "Non-Plan: An Experiment in Freedom," *New Society* 13, no. 338 (20 March, 1969): 435–43. For a reprint and commentaries, see Jonathan Hughes and Simon Sadler, eds., *Non-Plan: Essays on Freedom Participation and Change in Modern Architecture and Urbanism* (Oxford: Architectural Press, 2000).

15.   N. J. Habraken, *Supports* (1961; Urban International Press, 1999).

16.   Yona Friedman, *Toward a Scientific Architecture* (Cambridge: MIT Press, 1975).

17.   Hashim Sarkis, ed., *Le Corbusier's Venice Hospital and the Mat Building Revival* (New York: Prestel, 2001).

18.   Alex Gordon, "Architects and Resource Conservation," *RIBA Journal* (January 1974), pp. 9–12.

19.   Hans Moravec, *Mind Children: The Future of Robot and Human Intelligence* (Cambridge: Harvard University Press, 1988). For a return to the theme, see Hans Moravec, *Robot: Mere Machine to Transcendent Mind* (New York: Oxford University Press, 1999).

20.   For a stab at the brain science, see Joseph LeDoux, *Synaptic Self: How Our Brains Become Who We Are* (New York: Viking, 2002).

21.   As Mark C. Taylor has pointed out, such self-as-software speculations "revise ancient philosophical and theological visions for the twenty-first century." Proponents of this view are "contemporary Gnostics, Platonists, and Cartesians" who espouse "a thoroughgoing dualism between mind and body, form and matter, immateriality and materiality, pattern and substance, etc." *The Moment of Complexity: Emerging Network Culture* (Chicago: University of Chicago Press, 2002), p. 223.

22. The phrases are Ray Kurzweil's, from *The Age of Spiritual Machines: When Computers Exceed Human Intelligence* (New York: Viking, 1999).

23. There is an extensive literature of shedding flesh and virtual bodies. The fictional locus, in the 1980s and 1990s, was established by Vernor Vinge's novella *True Names* (1981; reprinted in James Frenkel, ed., *True Names and the Opening of the Cyberspace Frontier* [New York: Tor Books, 2001]) and the novels of William Gibson, particularly *Neuromancer* (New York: Ace Books, 1984). N. Katherine Hayles provides a critical introduction in *How We Became Posthuman: Virtual Bodies in Cybernetics, Literature, and Informatics* (Chicago: University of Chicago Press, 1999).

24. Ray Kurzweil has pursued this point: "There won't be mortality by the end of the twenty-first century. Not in the sense that we have known it. Not if you take advantage of the twenty-first century's brain-porting technology. Up until now, our mortality was tied to the longevity of our *hardware*. When the hardware crashed, that was it." *Age of Spiritual Machines*, pp. 128–29.

25. Freudians will be quick to point out one potentially good reason; you may not *like* your nature-and-nurture-given body very much, and Moravec's brain operation may belong in some murky category with cross-dressing, anorexia, body-piercing, and teenage suicide. But I shall not pursue this fascinating diversion here.

26. Here I paraphrase Bruno Latour. His short and provocative text, *We Have Never Been Modern* (Cambridge: Harvard University Press, 1993), was a witty riposte to the afflatus of postmodernism in French intellectual life.

CHAPTER 11 CYBORG AGONISTES

1. Construction of Palma Nova commenced in 1593. It was intended to serve as a fortified garrison outpost of Venice. The design is usually credited to the Venetian architect and urban theorist Vincenzo Scamozzi—author of the treatise *L'idea dell'architettura universale*, which deals extensively with fortified cities. Today, Palma Nova is a sleepy country town and one of the best surviving examples of what Lewis Mumford sardonically called the "asterisk plan."

2. Lewis Mumford, "Protection and Medieval Town," chapter 1 of *The Culture of Cities* (New York: Harcourt, Brace, 1938), pp. 13–64.

3. Lewis Mumford, *The City in History* (London: Secker, and Warburg, 1961), p. 410.

4. On the vulnerabilities of the oil and gas supply infrastructure—which, in the United States, consists of more than one million miles of natural gas pipeline and more than two hundred thousand miles of oil pipe—see Kathleen McFall, "Post-9/11 Investigations Reveal Oil, Gas Achilles Heel," in *Building for a Secure Future*, special editorial supplement to *Engineering News-Record* and *Architectural Record* magazines, Spring 2003, pp. 11–13. For a

discussion of the growing number of cyberattacks on the electrical grid, see Charles Piller, "Hackers Target Energy Industry," *Los Angeles Times*, 8 July 2002.

5. Joint Economic Committee, United States Congress, *Security in the Information Age: New Challenges, New Strategies*, May 2002, p. 2.

6. Ibid., p. 42.

7. Leonard Lee, "The Bigger They Are," in *The Day the Phones Stopped* (New York: Donald I. Fine, 1991), pp. 71–97.

8. Sandeep Junnarkar, "Keeping Networks Alive in New York," *CNET News.com*, 28 August 2002.

9. The story is recounted in Scott Charney, "Transition between Law Enforcement and National Defense," in Joint Economic Committee, United States Congress, *Security in the Information Age* (May 2002), pp. 52–60.

10. Paul Baran, *On Distributed Communications: 1. Introduction to Distributed Communications Network*, Memorandum RM-3420-PR (Santa Monica, Calif.: RAND Corporation 1964).

11. Bill Cheswick's beautiful, widely published Internet maps illustrate this. See <http://research.lumeta.com> (accessed December 2002). For a detailed discussion, with emphasis on reliability, see Albert-László Barabási and Eric Bonabeau, "Scale-Free Networks," *Scientific American* 288, no. 5 (May 2003): 60–69.

12. National Research Council, *The Internet under Crisis Conditions: Learning from September 11* (Washington, D.C.: National Academies Press, 2002).

13. The scale of attacks has escalated with that of the Internet itself. When Robert Morris's notorious worm struck in 1988, it crashed about 6,000 servers—roughly ten percent of the Internet's total at that time.

14. On the distribution of bioagents and the potential consequences, see Richard Preston, *The Demon in the Freezer* (New York: Random House, 2002).

15. On post-9/11 container security, see Aileen Cho, "Containing Container Risks and Connecting Modes," in *Building for a Secure Future*, special editorial supplement to *Engineering News-Record* and *Architectural Record* magazines, Spring 2003, pp. 19–20.

16. Martin Amis, "Fear and Loathing," *Guardian Unlimited*, 18 September 2001, <http://www.guardian.co.uk/Archive/Article/0,4273,4259170,00.html> (accessed December 2002).

17. Filters and barriers are relatively cheaper and easier to deploy in the digital world than the physical world, though advances in miniaturized, inexpensive sensor technology may begin to change this. It is likely that water, air, and other supply networks will increasingly be equipped with early warning systems that permit operators to swiftly quarantine affected sections.

18. For a more detailed discussion of this point see Thomas Homer-Dixon, "The Rise of Complex Terrorism," *Foreign Policy* (January/February 2002), <http://www.foreignpolicy.com/issue_janfeb_2002/homer-dixon.html> (accessed December 2002).

19. Charles V. Bagli and Leslie Eaton, "Seeking New Space, Companies Search Far From Wall St.," *New York Times*, 14 September 2001, pp. A1, A6.

20. William Safire, "An Optimist's What-if," *New York Times*, 29 October 2001, p. A15. He elaborated: "House members at home in their districts could be called into virtual session, as could the Senate; debates could be on the Internet, deals made in conference calls and votes taken (as they now are) electronically."

21. For a typical polemic in favor of increased decentralization, motivated by the World Trade Center attacks, see Oliver Morton, "Divided We Stand," *Wired*, December 2001, pp. 152–55. For some anecdotal indications of a shift in this direction, several months after September 11, see Charles V. Bagli, "Seeking Safety, Manhattan Firms Are Scattering," *New York Times*, 29 January 2002, pp. A1, A24. And for some analysis from an economic perspective, see Edward L. Glaeser and Jesse M. Shapiro, "Cities and Warfare: The Impact of Terrorism on Urban Form," Harvard Institute for Economic Research Discussion Paper 1942, December 2001, <http://papers.ssrn.com/abstract=293959> (accessed December 2002); and William C. Wheaton and Jim Costello, "The Future of Lower Manhattan, Signals from the Marketplace," MIT Center for Real Estate, 2002, <http://web.mit.edu/cre/www/news/ncnyc.html> (accessed December 2002).

22. Paul de Armond, "Netwar in the Emerald City: WTO Protest Strategy and Tactics," in John Arquilla and David Ronfeldt, eds., *Networks and Netwars* (Santa Monica, Calif.: RAND Corporation, 2001), pp. 201–35.

23. For a more detailed discussion of Critical Mass, with pointers to Web sites and newspaper accounts, see David Ronfeldt and John Arquilla, "What Next for Networks and Netwars?" in Arquilla and Ronfeldt, eds., *Networks and Netwars*, pp. 336–37.

24. See, for example, John Arquilla and David Ronfeldt, *Swarming and the Future of Conflict*, Document DB-311-OSD (Santa Monica, Calif.: RAND Corporation, 2000); and Bruce Berkowitz, *The New Face of War* (New York: Free Press, 2003). The idea of swarming was given considerable popular currency by Kevin Kelly, *Out of Control: The New Biology of Machines, Social Systems and the Economic World* (Reading, Mass.: Addison-Wesley, 1994).

25. Lakshmi Sandhana, "The Drone Armies Are Coming," *Wired News*, 30 August 2002.

26. "Lehman Brothers' Network Survives," *NetworkWorldFusion*, 26 November 2001.

27. "Cantor-Fitzgerald: 47 Hours," *Baseline*, <www.baselinemag.com/article2/0,3959,36807,00.asp> (accessed December 2002).

28. "Businesses Start the Recovery Process," *NetworkWorldFusion*, 12 September 2001.

29. See, for example, Sun and Nortel's Enterprise Continuity system, introduced in 2002.

30. Associated Press, "Powerful Attack Upsets Global Internet Traffic," *New York Times*, 23 October 2002, p. A19.

31. Rick Weiss, "Polio-Causing Virus Created in N.Y. Lab: Made-From-Scratch Pathogen Prompts Concerns About Bioethics, Terrorism," *Washington Post*, 12 July 2002, p. A01.

32. In his well-known article "Why the Future Doesn't Need Us," Bill Joy drew widespread attention to the dangers of runaway, destructive self-replication in genetics, nanotechnology, and robotics. As he remarked, "Gray goo would surely be a depressing ending to our human adventure on Earth, far worse than mere fire or ice, and one that could stem from a simple laboratory accident. Oops." *Wired* 8, no. 4 (April 2000). Note, however, that many scientists remain very skeptical about the possibility.

CHAPTER 12 LOGIC PRISONS

1. Much of the important early work was done at MIT's Project MAC (for Multiple Access Computer and Machine-Aided Cognition), which began in 1963. Project MAC's Multics multiuser timesharing system was a major landmark.

2. It is immediately evident that there will be a scaling problem; the size of the table grows geometrically as the numbers of users and numbers of files increase. Much of the practical technology of access control is focused upon efficiently handling large-scale problems.

3. Ann H. Crowe, Linda Sydney, Pat Bancroft, and Beverly Lawrence, *Offender Supervision with Electronic Technology* (Lexington, Ky: American Probation and Parole Association, 2002), p. 67. *Journal of Offender Monitoring* provides up-to-date information on electronic offender monitoring and its uses.

4. For a cogent summary and analysis of the related technical, legal, social, and political issues, see Stephen T. Kent and Lynette I. Millett, eds., *IDs—Not That Easy* (Washington, D.C.: National Academy Press, 2002).

5. The numerous complexities of identity verification and associated issues are discussed in compelling and often amusing detail in Bruce Schneier, *Secrets and Lies: Digital Security in a Networked World* (New York: John Wiley, 2000). For a more concise and anecdotal introduction, see Charles C. Mann, "Homeland Insecurity," *Atlantic Monthly* 290, no. 2 (September 2002): 81–102. On authentication technologies and their implications, see Stephen T. Kent and Lynette I. Millett, eds., *Who Goes There? Authentication through the Lens of Privacy* (Washington, D.C.: National Academies Press, 2003).

6. See, for example, Ravikanth Pappu, Neil Gershenfeld, Benjamin Recht, and Jason Taylor, "Physical One-Way Functions," *Science* 297, no. 5589 (September 20, 2002): 2026–30.

7. See the Web site of the Biometric Consortium, <www.biometrics.org> (accessed December 2002).

8. <www.passport.net> (accessed December 2002).

9. <www.projectliberty.org> (accessed December 2002).

10. National Electronic Commerce Coordinating Council (NECCC), *White Paper on Identity Management*, November 2002.

11. The case of DoubleClick Inc. brought the issue into sharp focus. DoubleClick used Web cookies, on a large scale, to track and profile visitors to Web sites and was sued by several states over its practices. See Robert O'Harrow Jr., "Web Ad Firm to Limit Use of Profiles," *Washington Post*, 27 August 2002, p. E01.

12. A typical Web site log file records the date, time, and IP address of a visitor, the site being accessed, the file downloaded and technical information about it, the browser type and operating system of the visitor, the link the visitor followed to get there, and if and when the visitor has been there before.

13. Elise Jordan and Arielle Levin Becker, "Princeton Officials Broke into Yale Online Admission Decisions," *Yale Daily News*, 25 July 2002, <www.yaledailynews.com> (accessed December 2002).

14. See, for example, Larry Ellison, "Digital IDs Can Help Prevent Terrorism," *Wall Street Journal*, 8 October 2001.

15. Techniques for this purpose go under the names of pattern recognition and classification, data mining, and knowledge discovery. They constitute a very large and active area of computer science and technology.

16. Robert O'Harrow Jr., "Air Security Focusing on Flier Screening," *Washington Post*, 4 September 2002.

17. Robert O'Harrow Jr., "Financial Database to Screen Accounts: Joint Effort Targets Suspicious Activities," *Washington Post*, 30 May 2002, p. E01, and Robert O'Harrow Jr., "In Terror War, Privacy vs. Security: Search for Illicit Activities Taps Confidential Financial Data," *Washington Post*, 3 June 2002, p. A01.

18. This point became vividly evident in the aftermath of the September 11 attacks. See, for example, Joel Garreau, "Disconnect the Dots," *Washington Post*, 17 September 2001, p. C01. The Web site of DARPA's Information Awareness Office, <www.darpa.mil/iao/> (accessed December 2002), comments: "The most serious asymmetric threat facing the United States is terrorism, a threat characterized by collections of people loosely organized in shadowy networks that are difficult to identify and define."

19. *Onion* 37, no. 34 (26 September 2001), <www.theonion.com/onion3734/us_vows_to_defeat_whoever.html> (accessed December 2002).

20. See John Sutherland, "No More Mr. Scrupulous Guy," *Guardian Unlimited*, 18 February 2002, <www.guardian.co.uk/Archive/Article/0,4273,4358017,00.html> (accessed December 2002); John Markoff, "Pentagon Plans a Computer System That Would Peek at Personal Data of Americans," *New York Times*, 9 November 2002, <www.nytimes.com/2002/11/09/politics/09COMP.html> (accessed December 2002); Robert

O'Harrow Jr., "U.S. Hopes to Check Computers Globally: System Would Be Used to Hunt Terrorists," *Washington Post*, 12 November 2002, p. A04; and William Safire, "You Are a Suspect," *New York Times*, 14 November 2002, p. A35. Poindexter ought to know about the use of databases to nail malefactors; key evidence used to convict him for his notorious Iran-Contra activities consisted of backup tapes of his email messages.

21. A March 2002 contractor solicitation (BAA 02-08) requested, "information technologies that will aid in the detection, classification, identification, and tracking of potential foreign terrorists, wherever they may be" based upon "gathering a much broader array of data than we do currently."

22. <www.darpa.mil/iao/> (accessed December 2002).

EPILOGUE

1. There have, of course, been many formulations of this principle. In *One World: The Ethics of Globalization* (New Haven: Yale University Press, 2002), pp. 168–69, Peter Singer cogently puts it as follows: "We might hold that we have a special obligation to our fellow citizens because we are all taking part in a collective enterprise of some sort. . . . It is therefore possible to see the obligation to assist one's fellow citizens ahead of citizens of other countries as an obligation of reciprocity, though one that is attenuated by the size of the community and lack of direct contact between, or even bare knowledge of, other members of the community." He quotes Walter Feinberg, *Common Schools/Uncommon Identities: National Unity and Cultural Difference* (New Haven: Yale University Press, 1998), p. 119: "The source of national identity is . . . connected to a web of mutual aid that extends back in time and creates future obligations and expectations."

2. Singer notes (*One World*, pp. 9–12): "For most of the eons of human existence, people living only short distances apart might as well, for all the difference they made to each other's lives, have been living in separate worlds. A river, a mountain range, a stretch of forest or desert, a sea—these were enough to cut people off from each other. Over the past few centuries the isolation has dwindled, slowly at first, then with increasing rapidity. Now people living on opposite sides of the world are linked in ways previously unimaginable. . . . This change creates the material basis for a new ethic that will serve the interests of all who live on this planet in a way that, despite much rhetoric, no previous ethic has ever done." On expanding moral circles in other contexts, see Peter Singer, *The Expanding Circle* (New York: Farrar, Straus and Giroux, 1981); Robert Wright, *NonZero* (New York: Pantheon, 2000); and Steven Pinker, *The Blank Slate* (New York: Viking, 2002), pp. 168, 320.

3. Henry Sidgwick, *The Methods of Ethics*, 7th ed. (London: Macmillan, 1907), p. 246. This passage is quoted, and taken as the starting point for extensive and insightful discussion, in Singer, *One World*, p. 153.

4. Plato, *Laws*, 737e ff.

5. Aristotle, *Politics*, 1326b II.

6. Ferdinand Tönnies, *Community and Society*, trans. Charles P. Loomis (1887; New York: Harper, 1963).

7. For an elaboration of this point, see Langdon Winner, "Complexity, Trust, and Terror," *NetFuture*, no. 137 (22 October 2002), <www.netfuture.org/2002/oct2202_137.html> (accessed December 2002).

8. Marx and Engels, of course, wrote in *The Communist Manifesto* (1848; London: Verso, 1998), p. 40, of how the growth of great industrial cities had "rescued a considerable part of the population from the idiocy of rural life." They were not just trash-talking the yokels, but, as Eric Hobsbawm points out in his introduction to this anniversary edition, referring to something akin to the Greek *idiotes*—that is, a person of "narrow horizons" or "isolation from the wider society" (p. 11).

9. Patrick E. Tyler, "A New Power in the Streets," *New York Times*, 17 February 2003, pp. A1, A8.

10. See, for example, Frances Cairncross, *The Death of Distance* (Boston: Harvard Business School Press, 1997).

ACKNOWLEDGMENTS

Me++ (2003) completes an informal trilogy that began with *City of Bits* (1995) and continued with *E-topia* (1999). These texts have resulted from a ten-year project of real-time scholarship that began with the explosive growth of the Internet and the World Wide Web in the mid-1990s, continued through the dot-com boom and bust that straddled the turn of the century, and has now extended into the era of wireless networks and embedded intelligence. My goal has been to identify emerging conditions, formulate crucial questions, suggest options, and stimulate critical discussion at a pace fast enough to make a difference.

Many of the themes and ideas that I have pursued have emerged from discussions with students and colleagues, and I am particularly grateful to those at MIT for their suggestions and stimulation. Anthony Townsend's research contributed substantially to chapter 11, "Cyborg Agonistes." Dan Greenwood, J. C. Hertz, Anthony Townsend, Harvey Waxman, Krzysztof Wodiczko, and Jane Wolfson all read drafts and made helpful suggestions. Several chapters appeared, in earlier forms, as papers and articles: chapter 6, "Digital Doublin'," appeared in *Archis*, no. 2 (May 2002): 22–31; chapter 9, "Post-Sedentary Space," was presented in 2002 at a Green College, University of British Columbia symposium on the city and was published as "What Cyberspace Does to Real Space," in *Topic Magazine*, no. 3 (2003): 34–41; chapter 11, "Cyborg Agonistes," was presented in 2002 at an MIT Department of Urban Studies and Planning symposium on "The Resilient City."

And most important has been the love and support of my family—Billy, Emily, and Jane.

INDEX

interim storage devices, 47
Internet, 10, 50, 140, 148, 149, 175, 177
Internet Archive, 37
Internet cafés, 155, 156
Iridium, 53, 54
Irish Times, 108

Jacobson, Joe, 136
Java, 141, 167
Jefferson, Thomas, 85
Joy, Bill, 186
Joyce, James, 19, 24, 85, 103–104, 107, 108–109, 112, 116, 145, 223n6, 233n1

Kahle, Brewster, 37
Kapuscinski, Ryszard, 63
KaZaA, 133, 233n26
Kenner, Hugh, 107, 109
King, Rodney, 206
Kircher, Athanasius, 25, 27
Kleinrock, Leonard, 57
Korea, 156
Kurzweil, Ray, 70, 243n24

land-use zoning, 162, 165
Las Vegas, 120
Latour, Bruno, 243n26
Laugier, Marc-Antoine, 24
Le Corbusier, 41
Lefebvre, Henri, 165, 215n6
L'Enfant, Pierre, 29, 174
LexisNexis, 87, 139
Liberty Alliance Project, 195
libraries, 85–86, 164
Library of Congress, 37, 88
Licklider, J. C. R., 34
light-emitting diode (LED) displays, 120–121, 155
lightness, 63
Linux, 141
Lisp, 140, 184

Lloyd, Seth, 38
location awareness, electronic, 114
location-based advertising, 145
location tracking, 115–116
logic prisons, 201
London, 10, 154, 174
Lonely Planet, 121
Loran, 116
Los Angeles, 10, 100, 206
low-earth-orbit (LEO) satellite systems, 53–54

Machover, Tod, 74
Magellan, Ferdinand, 203
Mann, Steve, 110
Mapquest, 122
maps, 121–122
Marconi, Guglielmo, 1–2, 48, 74, 210, 213n2
marketing, 150. *See also* electronic commerce
pinpoint marketing, 60
markets, 32–33, 59–60
Marx, Karl, 208, 249n8
Massachusetts Institute of Technology (MIT)
Athena network, 148, 155
Laboratory for Computer Science, 117
Media Laboratory, 49, 136
Massumi, Brian, 228n42
materiality, 3–4. *See also* dematerialized information; virtuality
Maxwell, James Clerk, 47
McLuhan, H. Marshall, 42, 53, 61, 214n3
medium-earth-orbit (MEO) satellite systems, 54
MEDLINE, 87
memory, 36–38
metadata, 123, 237n25

microelectromechanical systems
(MEMS), 30, 67, 69
microfabrication, 66–67, 68
Microsoft Corporation, 141, 142
Microsoft Passport, 195
mind, 34–36
miniaturization, 2, 46–47, 64–74,
76, 82, 83
mnemotechnics, 127–128
mobility, 57–58, 84
modernity, 31. *See also*
postmodernity
money, 93–96. *See also* banking
montage, 109
Moore, Charles W., 15–16
Moravec, Hans, 167
Morris, Robert, 244n13
Morris, William, 131
MP3 players, 64, 99, 134, 138, 140,
233n24
multifunctionality, 72, 76
Mumford, Lewis, 11, 169, 170,
243n1
music recording, 64, 81, 96–99, 138
and broadcasting, 98
digital, 98–99

nanobots, 70, 186
nanoelectromechanical systems, 69
nanotechnology, 67–69
Napster, 99, 139
National Academy of Sciences, 30
National Aeronautics and Space
Administration (NASA), 23, 53
Negri, Antonio, 215n6
networks, 9–11, 16, 17, 19
failure of, 171–174
New York City, 50, 152, 155, 158,
170, 174. *See also* World Trade
Center
NFL 2K3, 33
Nolli, Giambattista, 28
nomadicity, electronic, 57, 64, 159,

161. *See also* mobility
numerically controlled (NC)
production, 134

Oak Ridge National Laboratory, 30
object-oriented programming, 140
Odyssey, 1, 213n1, 233n1
Onion, 200
Orbcomm, 53

Packard, David, 86
paintings, 90–91
Palladio, Andrea, 99, 139
Palma Nova (Italy), 169, 170, 174,
175, 243n1
parallel processing, 13–14
ParcBIT, 152
Paris, 7, 29, 154, 174
passwords, 10, 193
pattern matching, 199
PayPal, 95
personal computers, 67
personal digital assistants (PDAs),
66, 72, 73, 77, 79
phenomenology, 39
photography, 65, 91–92, 139
digital, 65–66, 92
Piano, Renzo, 24
Plato, 205
plumbing, 22, 23–24, 219n15
Poindexter, John, 201, 248n20
points of presence, 143. *See also* fields
of presence
Polaroid Corporation, 65
Pompidou Center (Paris), 24
pornography, 146–147
postmodernity, 15–16
post-sedentary space, 59–61
power, electric, 20–21
power supply, 42, 46, 47
President's Critical Infrastructure
Protection Board, 10–11,
213n4